The Ca²⁺ Pump of Plasma Membranes

Authors:

Alcides F. Rega, Ph.D. Biochem. Pharm.

Professor of Biophysical Chemistry
Instituto de Química y Fisicoquímica Biologicas
Universidad de Buenos Aires
CONICET
Buenos Aires, Argentina

Patricio J. Garrahan, M.D.

Professor of Biophysical Chemistry
Instituto de Química y Fisicoquímica Biologicas
Universidad de Buenos Aires
CONICET
Buenos Aires, Argentina

CRC Press
Taylor & Francis Group
Boca Raton London New York

CRC Press is an imprint of the
Taylor & Francis Group, an **informa** business

First published 1986 by CRC Press
Taylor & Francis Group
6000 Broken Sound Parkway NW, Suite 300
Boca Raton, FL 33487-2742

Reissued 2018 by CRC Press

© 1986 by CRC Press, Inc.
CRC Press is an imprint of Taylor & Francis Group, an Informa business

No claim to original U.S. Government works

Library of Congress Cataloging in Publication Data

Rega. Alcides F.
    The Ca^{2+} pump of plasma membranes.

    Includes bibliographies and index.
    1. Plasma membranes. 2. Calcium. 3. Calcium in
the body. 4. Biological transport, Active.
I. Garrahan, Patricio J. II. Title. [DNLM:
1. Biological Transport.   2. Calcium--metabolism.
3. Cell Membrane--physiology. QV 276 R333c]
QH601.R44   1985   574.87'5   85-11707
ISBN 0-8493-6253-9

A Library of Congress record exists under LC control number: 85011707

Publisher's Note
The publisher has gone to great lengths to ensure the quality of this reprint but points out that some imperfections in the original copies may be apparent.

Disclaimer
The publisher has made every effort to trace copyright holders and welcomes correspondence from those they have been unable to contact.

ISBN 13: 978-1-315-89126-2 (hbk)
ISBN 13: 978-1-351-07036-2 (ebk)

Visit the Taylor & Francis Web site at http://www.taylorandfrancis.com and the
CRC Press Web site at http://www.crcpress.com

THE AUTHORS

Alcides F. Rega, Ph.D. Biochem. Pharm., is Director of the Department of Biological Chemistry and Professor of Biophysical Chemistry, Facultad de Farmacia y Bioquímica, Universidad de Buenos Aires (UBA), Argentina.

Dr. Rega received his undergraduate degree in Pharmacy and Biochemistry from the UBA in 1955. With a grant from the UBA, he pursued his postgraduate studies, obtaining his doctorate in 1962. He received postdoctoral training in the Department of Radiation Biology and Biophysics at the University of Rochester, New York, from 1964 to 1966.

Dr. Rega is an Established Investigator of the Consejo Nacional de Investigaciones Científicas y Técnicas (CONICET), Argentina, past president and founding member of the Sociedad Argentina de Biofísica and member of the International Cell Research Organization of UNESCO and the Commission on Education and Development in Biophysics of the International Union of Pure and Applied Biophysics.

His research interest is in transport ATPases. He has authored publications in international journals and in books on active transport of calcium and active transport of sodium and potassium.

Patricio J. Garrahan, M.D., is Professor of Biophysical Chemistry, Facultad de Farmacia y Bioquímica, Universidad de Buenos Aires, and a Principal Investigator of CONICET.

Dr. Garrahan received his medical degree from the UBA in 1960. With a fellowship from CONICET, he received postgraduate training as a research visitor in the Physiological Laboratory, University of Cambridge, England.

Dr. Garrahan is a member of the advisory committee on Medical Sciences of CONICET. He is also a member of the Commission on Cell and Membrane Biophysics and of the International Union of Pure and Applied Biophysics, a founding member and past president of the Sociedad Argentina de Biofísica, a founding member and member of the Council Academia de Ciencias de America Latina, and a member of the International Cell Research Organization of UNESCO.

Dr. Garrahan has published more than 50 papers in international journals or in chapters of books on active transport of calcium in red cells, partial reactions and elementary steps of ATP hydrolysis by the calcium pump of red cells, active transport of sodium and potassium in red cells, interaction of ligands during steady-state hydrolysis of ATP by the (Na, K)ATPase. His research interests are in the active transport of cations and transport of ATPases.

ACKNOWLEDGMENTS

We wish to express our gratitude to our colleagues H. Barrabin, A. J. Caride, R. B. Kratje, J. N. Larocca, H. Mugica, D. E. Richards, and J. P. F. C. Rossi for their experimental work on the Ca^{2+} pump performed in our laboratory and to all the scientists whose studies have made this book possible. The authors' research programs have been supported by the Consejo Nacional de Investigaciones Científicas y Técnicas (CONICET) of Argentina, the University of Buenos Aires, the Secretaría de Ciencia y Técnica of Argentina, the RLA 78/024 program from UNESCO and the Fundacion Roemmers.

TABLE OF CONTENTS

Chapter 1
The Cellular Calcium
A. F. Rega

I.   The Measurement of Ca²⁺ in the Cytosol..1
       A.   Total and Free-Ionic (Ca²⁺) Calcium 1
       B.   Methods to Determine Cytosolic Ca²⁺ Concentration 1
             1.   Ca²⁺-Binding Dyes 2
             2.   Ca²⁺-Activated Photoproteins..3
             3.   Ca²⁺-Selective Electrodes 3
             4.   Null Point Titration 5
II.   The Distribution of Calcium Among the Cell Components 6
III.   Calcium in the Cytosol 7
References 10

Chapter 2
The Ca²⁺ Homeostasis
A. F. Rega

I.   How Does Ca²⁺ Enter and Leave the Cell?...13
       A.   Excitable Cells..13
       B.   Nonexcitable Cells...14
II.   Transport of Ca²⁺ by Intracellular Organelles.......................................15
III.   The Role of the Transport Mechanisms in the Regulation of Ca²⁺
       Concentration in the Cytosol 16
References 19

Chapter 3
Calcium and Cell Function
P. J. Garrahan

I.   Introduction...21
II.   The Messenger Role of Ca²⁺...21
       A.   The Mechanism of the Increase in Cytosolic Ca²⁺ in Stimulated
             Cells...21
             1.   Ca²⁺ Channels in Excitable Membranes 22
             2.   Cell Membrane Phospho- and Polyphosphoinositides and
                   Receptor-Mediated Ca²⁺ Mobilization 24
             3.   The Release of Ca²⁺ from Intracellular Stores 24
III.   Ca²⁺-Binding Proteins..25
       A.   General Properties...25
       B.   Calmodulin..26
             1.   Distribution 27
             2.   Chemical and Physical Properties 27
             3.   Binding of Ca²⁺...27
IV.   Enzymes that Depend on Ca²⁺ and Calmodulin 29
       A.   Enzymes Involved in Glycogen Metabolism 29
             1.   Skeletal Muscle Phosphorylase Kinase.............................29
             2.   Glycogen Synthetase Kinase..30
             3.   Protein-Phosphatases...30

B.   Enzymes Involved in Cyclic Nucleotide Metabolism 31
C.   Enzymes Involved in Regulation of Muscle Contraction and Other
      Motile Processes 31
      1.   Myosin Light-Chain Kinase 31
      2.   Phospholamban Kinase 31
      3.   Protein Kinases of Sarcoplasmic Reticulum of Skeletal
            Muscle 31
D.   Nervous Tissue Protein Kinases 32
E.   Membrane Protein Kinases in Other Tissues 32
F.   Plasma Membrane Ca^{2+}-ATPases...32
G.   Other Enzymes 33
      1.   NAD Kinase..33
      2.   Platelet Phospholipase A2 33
      3.   15-Hydroxyprostaglandin Dehydrogenase 33
V. Ca^{2+}-Dependent Enzymes that Do Not Require Calmodulin 33
      A.   Protein Kinase C 33
      B.   Calpain and Calpastatin..34
VI.   Regulation of Cell Functions by Ca^{2+} 34
      A.   Endo- and Exocytosis...34
      B.   Cell Motility and Self-Assembling Cell Components.......................35
      C.   Cell Division 36
      D.   Membrane Channels 37
            1.   The Gárdos Effect 37
            2.   Intercellular Communication 38
References 39

Chapter 4
From the Discovery of the Ca^{2+} Pump in Plasma Membranes to the Demonstration of
  Its Ubiquitousness
A. F. Rega

I.   A Historical Review 45
      A.   The First Report 45
      B.   The Finding of the Ca^{2+} Pump in Plasma Membranes Other than
            the Erythrocyte 47
      C.   How to Identify a Ca^{2+}-Transporting System with the Ca^{2+} Pump
            from Plasma Membrane 49
II.   Some Properties of the Ca^{2+} Pump from Various Cell Types 51
      A.   Circulating Cells 52
      B.   Excitable Cells..52
      C.   Tissue Cells...53
      D.   Other Cells 54
      E.   Conclusion 55
References 56

Chapter 5
Isolation and Purification of the Ca^{2+} Pump
P. J. Garrahan

I.   The Main Difficulties and the First Attempts...59
II.   The Use of Calmodulin Affinity Chromatography to Purify the
      Ca^{2+}-ATPase 60

III.   Properties of the Purified Ca²⁺-ATPase...62
    A.   Stability...62
    B.   Molecular Weight and Composition 62
    C.   Kinetic Properties..63
    D.   Reconstitution of the Purified Enzyme 64
    E.   Immunological Reactivity..64
References 64

Chapter 6
Transport of Ca²⁺ and ATP Hydrolysis by the Ca²⁺ Pump
A. F. Rega

I.   Transport of Ca²⁺...67
    A.   Introduction...67
    B.   Preparations Used for Transport Studies 67
        1.   Intact Red Blood Cells...67
        2.   Resealed Ghosts 68
        3.   Inside Out Vesicles 69
        4.   Reconstituted Liposomes 69
        5.   Squid Axons 70
    C.   Dependence on Ca²⁺ Concentration 71
        1.   Activation by Ca²⁺...71
        2.   Inhibition by Ca²⁺...72
    D.   Substances and Treatments that Increase the Rate of Transport 74
    E.   The Electrical Balance During Transport of Ca²⁺ 75
        1.   Electrogenic Transport 75
        2.   Electroneutral Transport...76
II.   ATP Hydrolysis..77
    A.   Dependence on Ca²⁺ Concentration 77
        1.   Activation by Ca²⁺...77
        2.   Inhibition by Ca²⁺...78
        3.   The Mechanism of the Inhibition 79
III.   Dependence on ATP 80
    A.   Substrate Specificity...80
    B.   The Substrate Curve 81
    C.   Kinetic Analysis of the Substrate Curve.....................................83
        1.   Kinetic Schemes that Give Biphasic Substrate Curves............83
            a.   Two Different Enzymes 83
            b.   Two Active Sites in the Same Enzyme 83
            c.   The Substrate as Activator 84
        2.   Comparison of the Kinetic Equations.................................85
            a.   The Mathematical Equivalence of Rate Equations 85
    D.   On the State of ATP as the Substrate for the Overall Reaction 87
References 88

Chapter 7
Other Properties and Coupling of Ca²⁺ Transport and ATP Hydrolysis
A. F. Rega

I.   Other Properties 91
    A.   Specificity for Ca²⁺ 91

B.   The Apparent Affinity for Ca²⁺ 91
     1.   Modifiers of the Apparent Affinity for Ca²⁺ 93
     2.   The EGTA Effect...93
C.   The Number of Ca²⁺ Sites..95
D.   Dependence on pH 96
E.   Dependence on Temperature...97
II.   The Coupling Between Ca²⁺ Transport and ATP Hydrolysis....................99
A.   Energetics of Ca²⁺ Transport..99
B.   The Stoichiometry of Ca²⁺ Transport 99
C.   Reversal of the Ca²⁺ Pump 101
References 102

Chapter 8
Partial Reactions of the Ca²⁺ ATPase
P. J. Garrahan

I.   The Elementary Steps of ATP Hydrolysis 105
A.   Introduction.. 105
B.   Phosphorylation 105
     1.   Kinetics of the Phosphorylation Reaction 106
     2.   Reversal of Phosphorylation.. 107
     3.   Chemical Properties of the Phosphoenzyme 107
C.   Dephosphorylation 109
     1.   The $E_1 \sim P \rightleftharpoons E_2 \sim P$ Transition................................... 111
D.   The $E_2 \rightleftharpoons E_1$ Transition 113
II.   Reaction Scheme for the Hyrolysis of ATP 114
III.   Energy Changes During the Elementary Steps 115
IV.   The Phosphatase Activity of the Ca²⁺-ATPase 116
A.   General Properties... 116
B.   Kinetics.. 116
     1   The Substrate Curve 116
     2.   Dependence on Mg²⁺ 117
     3.   Activation by Ca²⁺... 118
     4.   Effects of Monovalent Cations 118
C.   The Interaction Between the Sites for pNPP and the Sites for
     ATP... 119
     1.   The High-Affinity Site.. 119
     2.   The Low-Affinity Site 119
     3.   ATP Hydrolysis During Phosphatase Activity.................... 119
D.   Phosphatase Activity and Active Ca²⁺ Transport 122
References 123

Chapter 9
Activation by Magnesium and by Alkali Metal Ions
P. J. Garrahan

I.   Magnesium 127
A.   Ca²⁺-ATPase Activities in the Absence of Added Mg²⁺ 127
B.   The Kinetics of Activation by Mg²⁺... 127
     1.   Activity vs. Mg²⁺ Concentration..................................... 128
     2.   Activity vs. MgATP Concentration 128
     3.   Activation by Mg²⁺ under Steady-State Conditions.............. 129

C.   The Relation between Ca^{2+}-ATPase Activity and the Concentration of Mg^{2+} 129

      1.   Activation by Mg^{2+} 130

      2.   Inhibition by Mg^{2+} 131

D.   The Mechanism of the Activation by Mg^{2+} 131

      1.   MgATP as the Substrate 131

      2.   Direct Binding of Mg^{2+} to the ATPase 132

II.   Alkali Metal Ions 132

A.   The Kinetics of Activation by Alkali Metal Ions 133

B.   The Sideness of Activation 134

C.   The Effects of Alkali Metal Ions on the Elementary Steps of the ATPase Reaction 134

D.   The Physiological Meaning of Activation.................................. 134

References 135

Chapter 10
Calmodulin and Other Physiological Regulators of the Ca^{2+} Pump
P. J. Garrahan

I.   Calmodulin.. 137

A.   Binding of Calmodulin to the Ca^{2+} ATPase 138

      1.   Role of Ca^{2+} in Calmodulin Binding 140

      2.   Extent of Calmodulin Dependence 141

      3.   Binding of Calmodulin under Physiological Conditions 141

B.   Effects of Calmodulin on the Steady-State Kinetics of the Ca^{2+}-ATPase 142

C.   Effects of Calmodulin on the Elementary Steps of the Ca^{2+}-ATPase 144

II.   Conditions and Treatments that Mimic the Effect of Calmodulin............. 144

A.   The Lipid Environment 145

B.   Proteolysis.. 146

C.   The Mechanism of the Calmodulin-Like Effects of Acidic Lipids and Proteolysis 147

III.   Other Physiological Regulators... 148

A.   Protein Activators and Inhibitors... 148

B.   Phosphoinositides 148

C.   Regulation by Phosphorylation 148

References 149

Chapter 11
Inhibitors of the Ca^{2+} Pump
P. J. Garrahan

I.   Introduction... 153

II.   Inorganic Ions 153

A.   Lanthanides 153

B.   Vanadate.. 154

III.   Calmodulin Antagonists 156

IV.   Compounds that React with Proteins 159

A.   N-Ethylmaleimide (NEM) 159

B.   Anion Channel (Band III) Inhibitors...................................... 161

C.   Fluorescein Derivatives.. 161

V.   Other Inhibitors 162
        A.   Quercetin 162
        B.   Ruthenium Red 162
References 162

Index 165

Chapter 1

THE CELLULAR CALCIUM

A. F. Rega

I. THE MEASUREMENT OF Ca^{2+} IN THE CYTOSOL

This subject has received much attention during the last few years as a consequence of the increased awareness of the crucial role of cytosolic Ca^{2+} as a second messenger in a number of processes. Ashley and Campbell[1] have edited a book on detection and measurement of free Ca^{2+} in cells, and an authoritative and up-to-date review of the subject by R. Y. Tsien[2] has been published recently.

A. Total and Free-Ionic (Ca^{2+}) Calcium

The actual amount of loosely bound or free-ionic calcium in the cytosol is much lower than the total amount of calcium, because most of the calcium in the cells is either sequestered by subcellular organelles or bound to cell structural components or molecules in the cytosol. To predict chemical reactions in which calcium participates, it is necessary to know the chemical potential of calcium which can be calculated knowing the concentration of free-ionic calcium under ideal conditions. The concentration of solutes in the cytosol makes it a nonideal solution. As a consequence of this, as with any other charged solute, free-ionic calcium will be exposed to electrical interactions with the charged components in the cytosol so that the concentration of free-ionic calcium will be lower than that of free calcium. There are two ways to overcome this difficulty: one is to use correction factors[3] (activity coefficients) which account for the decrease in concentration or in chemical potential due to electrical interactions, and the other is to use Ca^{2+}-selective electrodes which allow one to measure the actual amount of ionized calcium, because they respond to free-ionic calcium concentrations. The use of correction factors presents a difficulty in that they vary in value from author to author and are not reliable for complex mixtures. Anyway, measurements with electrodes have always had to be made relative to standard solutions to which numerical values of free-ionic calcium concentration have been assigned. From a practical point of view, this means that (as has been clearly stated by Tsien[2]) "when one says, for example, that the free-ionic calcium concentration in a cell was measured to be 1 μM, that really means that the calcium activity in the cell was the same as the calcium activity in a certain calibrating solution in which the other major ionic constituents were considered to be similar to those of cytosol and which contained 1 μM of calcium ions not tightly bound to ligands." This definition is also valid for all other techniques used to measure free-ionic calcium concentration. Throughout this book, Ca^{2+} will stand for free-ionic calcium.

B. Methods to Determine Cytosolic Ca^{2+} Concentration

These methods were not available until a few years ago because of the difficulties emanating from the fact that: (1) cytosolic Ca^{2+} is always in the 10^{-6} to 10^{-9} M range and has to be measured in the presence of levels of Mg^{2+}, Na^+, and K^+ which are orders of magnitude higher; (2) the method to determine cytosolic Ca^{2+} concentration must allow the insertion of the Ca^{2+} probe into the cell without causing any damage to the structure and function of the cell; and (3) the Ca^{2+} probe must not bind significant amounts of Ca^{2+} from the cytosol.

Four methods are now available to measure the concentration of intracellular Ca^{2+}:

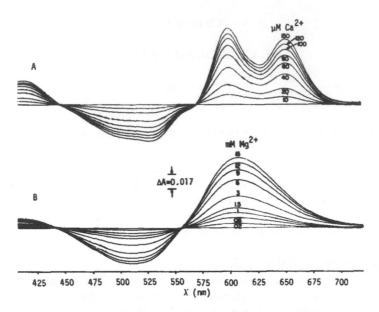

FIGURE 1.   Differential spectrum of arsenazo III vs. arsenazo III and various concentrations of Ca²⁺ (A) and Mg²⁺ (B). (From Scarpa, A., *Detection and Measurement of Free Ca²⁺ in Cells,* Ashley, C. C. and Campbell, A. K., Eds., Elsevier/North-Holland, Amsterdam, 1979, 96. With permission.)

(1) Ca²⁺-binding dyes; (2) Ca²⁺-activated photoproteins; (3) Ca²⁺-selective electrodes; and (4) null point titration.

1. Ca²⁺-Binding Dyes

There are two main classes of substances that undergo changes in their optical spectra upon Ca²⁺ binding: metallochromic indicators and tetracarboxylate dyes. Arsenazo III and antipyrylazo III are the most common metallochromic indicators.[4] Both are derivatives of 2,7-bisazo-1,8-dihydroxy-3,6-naphthalenedisulfonic acid. Ca²⁺ binds to arsenazo III with a K_d near 60 μM and increases arsenazo absorbance at 595 and 658 nm (Figure 1). The change can be detected measuring the differential spectrum of free arsenazo III vs. arsenazo III plus Ca²⁺, and it is not a linear function of Ca²⁺. Mg²⁺ produces a single broader change in absorbance with a maximum at 608 nm (Figure 1) so that, although arsenazo III is not specific for Ca²⁺, it can be made selective through the use of an appropriate pair of wavelengths. Combination of Ca²⁺ with antipyrylazo III increases the absorbance of the dye in the red region of the visible spectrum, while Mg²⁺ produces no change in this region. This property of antipyrylazo III makes it suitable for measuring Ca²⁺ concentrations without the interference of Mg²⁺ (but not Sr²⁺) by differential absorbance at 720 to 790 nm. The K_d of antipyrylazo III for Ca²⁺ is higher than that of arsenazo III and lies between 60 and 500 μM. At Ca²⁺ concentrations near to or greater than the K_d, the changes in absorbance are nonlinear with Ca²⁺ concentration.

Because of the relatively low differential extinction coefficient between the free-indicator and the calcium-indicator complexes, considerable amounts of the indicators are needed for the assay. This may cause undesirable changes in the concentration of Ca²⁺ in the presence of the indicators. Another problem is that the stoichiometry of the dye-calcium complex changes form 1:1 to 2:1 at concentrations of either dye greater than 1 μM. Metallochromic indicators do not permeate and have to be injected into the cells. Their use for measuring cytosolic Ca²⁺, therefore, is limited to large cells. One major advantage of these Ca²⁺ indicators is their fast response time which becomes important when changes in Ca²⁺ concentration are to be measured.

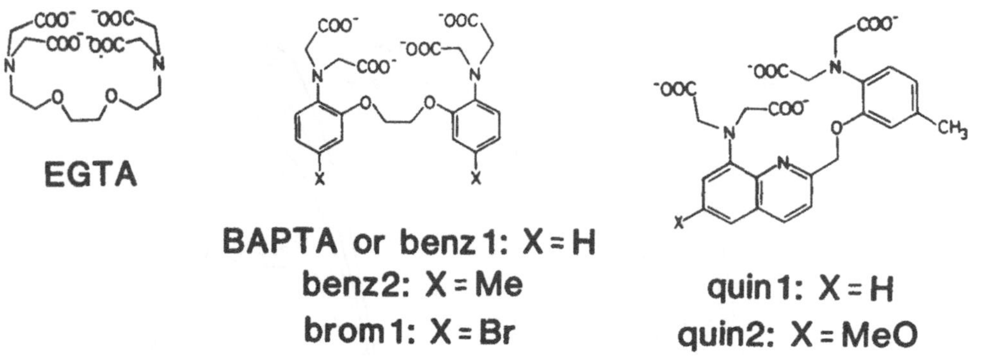

FIGURE 2. Structure of EGTA and some carboxylate dyes. (From Tsien, R. Y., *Ann. Rev. Biophys. Bioeng.*, 12, 91, 1983. With permission.)

Recently a new series of tetracarboxylate dyes (Figure 2) derived from 1,2-bis(*o*-aminophenoxy)ethane-*N,N,N',N'*-tetracetic acid (BAPTA), a molecule very close to EGTA, or from 2,2-bis(ethoxycarboxyl)methylamino-5-methylphenoxymethyl-6-methoxy-8-bis-(ethoxycarbonyl)methylaminoquinoline (quin2), have been introduced as Ca^{2+} indicators.[5,6] These chelators show high affinity (K_d 0.11 and 0.08 μM for BAPTA and quin2, respectively), 1:1 stoichiometry for Ca^{2+}, and high selectivity for Ca^{2+} over Mg^{2+}. Their absorbance and fluorescence increase largely upon Ca^{2+}-binding in a way which is not directly proportional to Ca^{2+} concentration (Figure 3). These tetracarboxylic acids are hydrophilic and do not permeate the membrane except when masked with esterifying groups. The resulting esters are hydrophobic and readily diffuse into all cells of a given population. Cytoplasmic esterases then hydrolyze the ester groups and restore the parent tetracarboxylic dye. This method of entrapping the tetracarboxylate chelators is particularly applicable to small cells in suspension.[2,6] This, together with their negligible binding to biological material, makes either BAPTA or quin2 a highly satisfactory means for nondestructively measuring Ca^{2+} concentrations in the cytosol.

2. Ca^{2+}-Activated Photoproteins

Aequorin is a protein of M_r near 31,000 (isolated from the jellyfish *Aequorea foskalea*) which, in the excited state, reacts with Ca^{2+} to emit one photon of light of about 460 nm wavelength. It combines with two Ca^{2+} with an apparent K_d of the same order as those of the metallochromic indicators. However, in an ionic environment similar to that within the cell, the apparent K_d increases to 10^{-3} to 10^{-4} M. This is fortunate, because at Ca^{2+} concentrations of about 0.1 μM, the percentage of aequorin bound to Ca^{2+} will be low, it will be consumed at a low rate, and no disturbances in cytoplasmic Ca^{2+} concentration will occur. Aequorin is highly selective for Ca^{2+}. Mg^{2+} retards the rate of light emission and at physiological concentrations lowers the sensitivity of aequorin to pH changes. A problem with aequorin is that its light output is an exponential function of Ca^{2+} concentration (Figure 4). A relation of this type complicates attempts at quantification of transient changes in light output, because the same increment in total Ca^{2+} will give different increments in light if it is confined to a small volume or distributed evenly throughout a cell. Another drawback to the use of aequorin is that it has to be introduced into cells by microinjection or reversible lysis.

3. Ca^{2+}-Selective Electrodes

The simplest method for the measurement of Ca^{2+} concentration is the Ca^{2+}-selective electrode (Figure 5). The main advantage of selective electrodes is that they measure

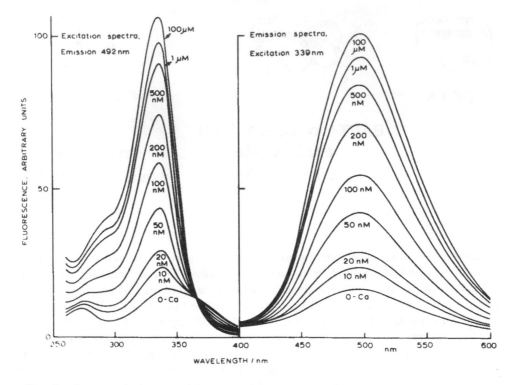

FIGURE 3.   Excitation and emission spectra of 20 μM quin2 with varying Ca²⁺ concentration in a solution containing 120 to 135 mM K⁺, 20 mM Na⁺, 1 mM Mg²⁺, pH 7.05. (From Tsien, R. Y., Pozzan, T., and Rink, T. J., *J. Cell. Biol.*, 94, 325, 1982. With permission.)

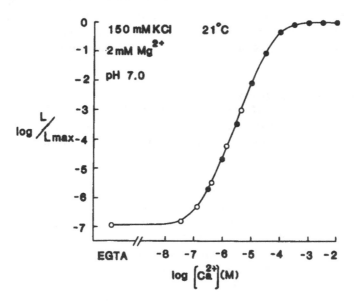

FIGURE 4.   The relation between the fractional light emission and Ca²⁺ concentration for aequorin. Filled circles, Ca²⁺ concentration established by dilution of 1 M CaCl₂; blank circles, Ca²⁺ concentration established by CaEGTA buffer mixtures. (From Allen, D. G. and Blinks, J. R., *Detection and Measurement of Free Ca²⁺ in Cells*, Ashley, C. C. and Campbell, A. K., Eds., Elsevier/North-Holland, Amsterdam, 1979, 96. With permission.)

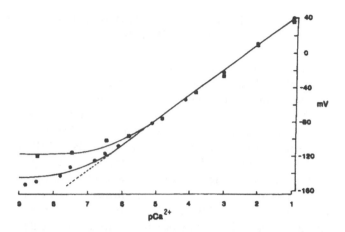

FIGURE 5.   Calibration curve of a Ca²⁺-selective electrode. Circles denote solution containing 200 m*M* K⁺ and squares denote solutions containing 200 m*M* K⁺ and 1 m*M* Mg²⁺.(From Owen, J. D. and Mack Brown, H., *Detection and Measurement of Free Ca²⁺ in Cells*, Ashley, C. C. and Campbell, A. K., Eds., Elsevier/North-Holland, Amsterdam, 1979, 96. With permission.)

the actual Ca^{2+} concentration without disturbing the equilibrium between bound and free calcium that the indicators that react with Ca^{2+} could alter.

The performance of ion-selective electrodes is highly dependent on the composition of the membrane in the electrode.[7] The use of electrically neutral ionophores for the membrane has made available Ca^{2+}-selective electrodes which have a detection limit of 0.1 μM or less and a selectivity for Ca^{2+} so high that it makes them almost immune from Mg^{2+} and other ionic species within cells.[7] There are, however, some difficulties in the use of electrodes for measuring the concentration of Ca^{2+} in the cytosol. The cell has to be punctured to allow the tip of the electrode to enter the cell, so that the method could damage the plasma membrane of the cell and cause leakage. On the other hand, because of the tip size, electrodes cannot be used in small cells. Another difficulty with electrodes is their low response speed which makes them unsuitable for measuring transients in Ca^{2+} concentration. Microelectrodes measure the local Ca^{2+} concentration in the surroundings of their tip. This is advantageous when point sampling within the cytoplasm of a cell is desired, but it represents a major drawback to the use of electrodes for measuring the mean Ca^{2+} concentration in the cytoplasm.

4. Null Point Titration

This method consists of measuring the change in Ca^{2+} concentration of the suspending medium due to penetration of Ca^{2+} after the plasma membrane of the cells has been made permeable to small solutes.[8] Digitonin is generally used to render the membrane permeable to Ca^{2+}, because it is effective on cholesterol-rich membranes, but has little effect on mitochondrial membranes. The Ca^{2+} of the suspending medium can be measured with a Ca^{2+}-selective electrode. A series of measurements at various initial extracellular Ca^{2+} concentrations are made, and the change in extracellular Ca^{2+} concentration is plotted as a function of the initial extracellular Ca^{2+} to obtain a straight line. The intercept of the line with the abscissa gives a null point when there is no net movement of Ca^{2+} into or out of the cells after the plasma membrane has been made permeable. If there is no membrane potential, the null point should correspond to the Ca^{2+} in the cytosol. Null point titration cannot be used for determining transient changes in Ca^{2+} concentration.

II. THE DISTRIBUTION OF CALCIUM AMONG THE CELL COMPONENTS

Total calcium in cells is represented by bound and free calcium, a large portion of the latter being Ca^{2+}. Bound calcium is combined mainly to anionic sites of sugars, proteins, phospholipids, phosphate, and phosphate esters. The concentration of calcium goes from 0.02 mmol/ℓ cell in cells containing no subcellular organelles up to about 2 mmol/ℓ cell in those cells that contain organelles. Measuring cellular calcium and fractionating it among the different cell components is a difficult task, because cellular components can be damaged during fractionation, changing the concentration of calcium they had originally in the intact cell. Few studies on the matter have been reported. Electron microprobe X-ray analysis, for instance, represents a nondestructive method of high specificity for calcium which, provided the biological sample is not altered by the preparation procedure, may be suitable to measure calcium distribution. The method, however, is complex and expensive.

Murphy et al.[8] applied a rapid cell fractionation technique to fractionate the calcium content of isolated hepatocytes from rat liver among the cellular subfractions. They found that the cells contained a total of 5.01 μmol calcium/g cell protein. After fractionation, 60 to 80% was isolated with the mitochondria and of it, only about 0.6% was Ca^{2+}. Most of the remainder was found associated to a microsomal fraction which may represent the endoplasmic reticulum. Murphy et al.[8] also found that the distribution of calcium mentioned above was independent of total cellular calcium between 3.2 to 31 μmol/g cell protein. Essentially similar results have been reported by Foden and Randle.[9] These results show that in hepatocytes, intracellular organelles are the main reservoir of calcium.

Blaustein et al.[10] have reported that total calcium content of presynaptic nerve terminals at rest isolated from rat brain is 5 to 10 μmol/g protein. About 70% of it remains associated to intracellular organelles after lysis, and 10% may represent calcium in the axoplasm; 5 to 10% of calcium in the axoplasm may be bound to proteins. Of the calcium in the organelles, about 50 to 75% is associated with mitochondria and can be released with uncouplers of oxidative phosphorylation like FCCP, and 20 to 25% is probably associated with components of the smooth endoplasmic reticulum and can be released with the calcium ionophore A23187.

Results of electron probe analysis in cryosections of skeletal and smooth muscle[11] suggest that much of intracellular calcium is accumulated within the sarcoplasmic reticulum rather than the mitochondria. This assertion is supported by the finding that: (1) electron probe analysis in skeletal muscle cells reveals that the terminal cisternae of sarcoplasmic reticulum contains 66 mmol total calcium/kg dry weight, whereas total calcium in the cytosol is 1 mmol/kg dry weight, and (2) mitochondria from smooth muscle cells do not appear loaded with calcium even after the cytoplasmic Ca^{2+} concentration is maintained at 1 μM for 30 min. It seems reasonable to think that the abundance of sarcoplasmic reticulum and its high affinity for calcium will lower the concentration of calcium in the cytosol of muscle cells below the level where effective Ca^{2+} uptake by mitochondria, whose affinity for Ca^{2+} uptake is orders of magnitude lower than that of the sarcoplasmic reticulum, is feasible. In light of the results mentioned above, it seems that in some specialized cells, the role of mitochondria in calcium storage might not be important.

The red blood cell can be taken as an example of a cell without intracellular organelles. The total content and the distribution of calcium in red blood cells have been studied by Harrison and Long.[12] They washed human and red blood cells with isotonic sodium chloride and after dry-ashing, measured calcium by atomic absorption spectrometry. The mean value of total calcium they found for normal subjects was 0.015

mmol/l cells, that is, about 100 times less than that in organelle-containing cells. These authors also observed that the bulk of the erythrocyte calcium is bound to the membrane, since after lysis of the cells, most of it was recovered with the membranes. Furthermore, about 90% of the calcium was removed by washing the cells with *isotonic sodium chloride containing 5 mM EGTA as a calcium complexing agent showing that*, although there are also binding sites for calcium on the internal surface (see below), the largest portion of calcium in the circulating erythrocyte is bound to the external surface of the membrane.

Long and Mouat[13] reported that the Scatchard plot of calcium uptake at various Ca^{2+} concentrations in the suspending media demonstrate the presence of three classes of binding sites for calcium on the external surface of the membrane of human red blood cells. *Sialic acid from the glycoproteins almost surely accounts for one of the site class population* (see also References 14 and 15), because there exists an approximate correlation between the capacity of calcium binding and the sialic acid content of erythrocytes that have been *partially depleted of the acid by treatment with neuraminidase*. After complete removal of sialic acid, red blood cells still retain about 50% of the amount of calcium found in the intact cell. Long and Mouat[13] also found that the amount of calcium that isolated membranes from red blood cells can bind is twice as much as that found in intact red blood cells. Since the difference between membranes and intact cells is that in the former, calcium from the suspending solution has access to both surfaces of the membrane, that finding was taken as an indication that, apart from those of the Ca^{2+}-pump, there are also sites that bind calcium on the internal surface of the red blood cell membrane.

In summary, although it is difficult to assess the distribution of calcium among the different cell components, it seems that in excitable as well as in nonexcitable cells, calcium associated with the mitochondria, the endoplasmic and sarcoplasmic reticulum, and plasma membrane accounts for most of the cell calcium. Table 1 summarizes the data presented here.

III. CALCIUM IN THE CYTOSOL

As in the entire cell, calcium in the cytosol is represented by bound and free calcium, the former being bound to soluble ligands in this particular case. Since human red blood cells do not contain subcellular organelles, no contribution of calcium leaking from the organelles to calcium in the cytosol can be expected. Data in Table 1 show that total calcium concentration in the cytosol of human red blood cells, estimated as *the difference between total calcium and calcium bound to the membrane, is low*. Out of this, only a fraction is Ca^{2+}.

Evidence has been collected showing that potassium ions diminish binding of calcium to red blood cells. Other monovalent cations behave like potassium also. This effect could be due to either shielding of the negative charges of the binding sites; lowering of the calcium activity by the cations; or both. A large fraction of the calcium binding sites do not discriminate well among divalent cations. Magnesium is by far the more abundant divalent cation in the cytosol of mammalian cells, its concentration being at least 100 times higher than that of calcium. Hence, competition of Mg^{2+} at the binding sites for Ca^{2+} may be an additional factor contributing to lowering the amount of bound calcium in the cytosol.

It is now accepted that under resting conditions, most cells contain very low concentration of Ca^{2+} in the cytosol. This fact has been confirmed in every cell in which Ca^{2+} has been measured. Table 2 gives values of the concentration of Ca^{2+} in the cytosol of a variety of cells together with the method used for the measurement. Regardless of the cell class and the procedure used for detection, all the values for cytosolic Ca^{2+} concentration are below 10^{-6} M.

Table 1

DISTRIBUTION OF CALCIUM AMONG THE COMPONENTS OF DIFFERENT CELLS AT REST[a]

	Hepatocyte[9]		Presynaptic nerve terminal at rest[10]		Striated muscle cell[11]		Human red blood cell[12]
	μmol/g cell protein	mmol/l cell[b]	μmol/g synaptosome protein	mmol/l cell[b]	μmol/g dry wt.	mmol/l cell[c]	(mmol/l cell)
Intact cell	5.01	1.00	5—10	1—2	—	—	0.015
Mitochondria	3.77	0.75	5	1	—	—	—
Endoplasmic reticulum	1.24	0.25	1—2	0.2—0.4	—	—	—
Sarcoplasmic reticulum (terminal cisternae)	—	—	—	—	66	0.94	—
Plasma membrane	—	—	—	—	—	—	0.013
Cytosol	—	—	—	—	1	0.28	—

a   The figures represent total calcium.
b   Calculated assuming that 1 g of protein is equivalent to 5 ml of cells and synaptosomes.
c   Calculated assuming that 1 g of dry weight is equivalent to 3.5 ml of cells and that the volume of sarcoplasmic reticulum represents 5 % of the cell volume.

Table 2
THE CONCENTRATION OF Ca²⁺ IN THE CYTOSOL

Tissue	Ca²⁺ concentration (μM)	Method[a]	Comment	Ref.
Squid giant axon[b] (*Loligo forbesi*)	About 0.3, (about the same as in a mixture of 45 mM CaEGTA:55 mM free EGTA)	1	All the intracellular Mg (10 mM) is assumed to be ionized	16
Squid giant axon[b] (*Loligo forbesi*)	About 0.1	1	The ionized Mg^{2+} is 3—4 mM	17
Squid giant axon[b] (*Loligo pealei*)	0.02	1		18
	0.05	2		
Squid giant axon[b] (*Loligo pealei*)	0.040	2		19
Squid giant axon[b] (*Doryteuthis plei*)	0.106	2		20
Giant axon of the[b] worm (*Myxicola infundibulum*)	0.1	1		21
Giant neurons of *Aplysia*	0.54	3		22
Neurons of *Helix aspersa*	0.17	3		23
Frog skeletal muscle	Below 0.2	1	No light was detectable upon injection of aequorin to resting muscle	24
Barnacle skeletal muscle	0.07—0.10	1		25
Tibialis anterior muscle of *R. temporaria*	Below detection limit	1		26
Heart papilliary muscle of ferret	0.26 at rest; during contraction raises up to 10	3		27
Rat heart muscle	0.270	7		28
Isolated smooth muscle cells from the stomach of the toad *Bufo marinus*	Below detection limit	1		29
Epithelial cells of *Necturus maculosus* kidney	0.34	3		30
Sheep cardiac Purkinje fiber	0.10—0.26	3		31
Sheep cardiac Purkinje fiber	0.075—0.590	3		32
Rat isolated hepatocytes	0.19	2	The value is the [Ca_i^{2+}] in equilibrium with the [Ca_e^{2+}] in the absence of membrane potential	33
Human red blood cells	0.18	3	The value is the activity of Ca^{2+} in the hemolyzate after addition of digitonin	34
Human red blood cells	0.028 $\mu M/l$ cells	6	Calculated from the total cytoplasmic calcium measured after loading the cell with a calcium chelator	35
Pigeon red blood cells	0.28	3	Calculated from the total cytoplasmic calcium measured after loading the cell with a calcium chelator	36

Table 2 (continued)
THE CONCENTRATION OF Ca²⁺ IN THE CYTOSOL

Tissue	Ca²⁺ concentration (μM)	Method[a]	Comment	Ref.
Human platelets	0.1	4		37
Pig lymphocytes	0.123	4		39
Amoeba (*Chaos carolinensis*)	0.05—0.1	1		39
Xenopus laevis embryos	0.32	3		40
Unfertilized mouse oocytes	0.05	1		41

[a] 1 = Recording the light produced by injected aequorin
2 = Differential absorbance of the Ca²⁺ complexed by arsenazo III
3 = Ca²⁺-selective electrode
4 = Recording the fluorescence produced by quin2
5 = Atomic absorption
6 = Isotopic dilution of ⁴⁵Ca²⁺
7 = Null point titration

[b] The calcium content of squid axons depends very much upon the conditions in which they were stored after isolation.

REFERENCES

1. **Ashley, C. C. and Campbell, A. K.**, Eds., *Detection and Measurement of Free Ca²⁺ in Cells,* Elsevier/North-Holland, Amsterdam, 1979.
2. **Tsien, R. Y.**, Intracellular measurements of ion activities, *Annu. Rev. Biophys. Bioeng.,* 12, 91, 1983.
3. **Vericat, F. and Grigera, J. R.**, Theoretical single-ion activity of calcium and magnesium ions in aqueous electrolyte mixtures, *J. Phys. Chem.,* 86, 1030, 1982.
4. **Scarpa, A.**, Measurement of calcium ion concentrations with metallochronic indicators, in *Detection and Measurement of Free Ca²⁺ in Cells,* Ashley, C. C. and Campbell, A. K., Eds., Elsevier/North-Holland, Amsterdam, 1979, 85.
5. **Tsien, R. Y.**, New calcium indicators and buffers with high selectivity against magnesium and protons: design, synthesis and properties of prototype structures, *Biochemistry,* 19, 2396, 1980.
6. **Tsien, R. Y., Pozzan, T., and Rink, T. J.**, Measuring and manipulating cytosolic Ca²⁺ with trapped indicators, *TIBS,* 9, 263, 1984.
7. **Ammann, D., Meier, P. C., and Simon, W.**, Design and use of calcium-selective microelectrodes, in *Detection and Measurement of Free Ca²⁺ in Cells,* Ashley, C. C. and Campbell, A. K., Eds., Elsevier/North-Holland, Amsterdam, 1979, 117.
8. **Murphy, E., Coll, K., Rich, T. L., and Williamson, J. R.**, Hormonal effects on calcium homeostasis in isolated hepatocytes, *J. Biol. Chem.,* 255, 6600, 1980.
9. **Foden, S. and Randle, P. J.**, Calcium metabolism in rat hepatocytes, *Biochem. J.,* 170, 615, 1978.
10. **Blaustein, M. P., Ratzlaff, R. W., and Kendrick, N. K.**, The regulation of intracellular calcium in presynaptic nerve terminals, *Ann. N.Y. Acad. Sci.,* 307, 195, 1978.
11. **Somlyo, A. P., Somlyo, A. V., Shuman, H., Sloane, B., and Scarpa, A.**, Electron probe analysis of calcium compartments in cryo sections of smooth and striated muscles, *Ann. N.Y. Acad. Sci.,* 307, 523, 1978.
12. **Harrison, D. G. and Long, C.**, The calcium content of human erythrocytes, *J. Physiol.,* 199, 367, 1968.
13. **Long, B. and Mouat, B.**, The binding of calcium ions by erythrocytes and ghost-cell membranes, *Biochem. J.,* 123, 829, 1971.
14. **Jacques, L. W., Brown, E. B., Barret, J. M., Brey, W. S., Jr., and Weltner, W., Jr.**, Sialic acid, calcium-binding carbohydrate, *J. Biol. Chem.,* 252, 4533.
15. **Porzig, H. and Stoffel, D.**, Equilibrium binding of calcium to fragmented human red blood cell membranes and its relation to calcium-mediated effects on cation permeability, *J. Membr. Biol.,* 40, 117, 1978.

16. Baker, P. F., Hodgkin, A. L., and Ridway, E. B., Depolarization and calcium entry in squid giant axons, *J. Physiol.*, 218, 709, 1971.
17. Baker, P. F., The use of aequorin in giant axons, in *Detection and Measurement of Free Ca²⁺ in Cells*, Ashley, C. C. and Campbell, A. K., Eds., Elsevier/North Holland, Amsterdam, 1979, 175.
18. DiPolo, R., Requena, J., Brinkley, F. J., Mullins, L. J., Scarpa, A., and Tiffert, J. T., Ionized calcium concentrations in squid axons, *J. Gen. Physiol.*, 67, 433, 1976.
19. Brinkley, F. J., Techniques for measuring free calcium *in situ* in single isolated cells using aequorin and metallochromic indicators, in *Detection and Measurement of Free Ca²⁺ in Cells*, Ashley, C. C. and Campbell, A. K., Eds., Elsevier/North Holland, Amsterdam, 1979, 319.
20. DiPolo, R., Rojas, H., Bergara, J., López, R., and Caputo, C., Measurements of intracellular ionized calcium in squid giant axons using calcium-selective electrodes, *Biochim. Biophys. Acta*, 728, 311, 1983.
21. Baker, P. F. and Schlaepfer, W. W., Uptake and binding of calcium by axoplasm isolated from giant axons of *Loligo* and *Myxicola*, *J. Physiol.*, 276, 103, 1978.
22. Owen, J. D. and Mack Brown, H., Comparison between free calcium selective microelectrodes and arsenazo III, in *Detection and Measurement of Free Ca²⁺ in Cells*, Ashley, C. C. and Campbell, A. K., Eds., Elsevier/North Holland, Amsterdam, 1979, 396.
23. Alvarez-Leefmans, F. J., Rink, T. J., and Tsien, R. Y., Free calcium ions in neurons of *Helix aspersa* measured with ion-selective micro-electrodes, *J. Physiol.*, 315, 531, 1981.
24. Blinks, J. R., Rudel, R., and Taylor, S. R., Calcium transients in isolated amphibian skeletal muscle fibers: detection with aequorin, *J. Physiol.*, 277, 291, 1978.
25. Campbell, A. K., Lea, T. J., and Ashley, C. C., Coelenterate photoproteins, in *Detection and Measurement of Free Ca²⁺ in Cells*, Ashley, C. C. and Campbell, A. K., Eds., Elsevier/North Holland, Amsterdam, 1979, 13.
26. Taylor, S. R., Rudel, R., and Blinks, R. R., Calcium transients in amphibian muscle, *Fed. Proc. Fed. Am. Soc. Exp. Biol.*, 34, 1379, 1975.
27. Marban, E., Rink, J. T., Tsien, R. W., and Tsien, R. Y., Free calcium in heart muscle at rest and during contraction measured with Ca²⁺-sensitive microelectrodes, *Nature*, 286, 845, 1980.
28. Williamson, J. R., Williams, R. J., Coll, K. E., and Thomas, A. P., Cytosolic free Ca²⁺ concentration and intracellular calcium distribution of Ca²⁺-tolerant isolated heart cells, *J. Biol. Chem.*, 258, 13411, 1983.
29. Fay, F. S., Shlevin, H. H., Granger, W. C., and Taylor, S. R., Aequorin luminiscence during activation of single isolated smooth muscle cells, *Nature*, 280, 506, 1979.
30. Lee, C. O., Taylos, A., and Windkager, E. E., Cytosolic calcium ion activity in epithelial cells of *Necturus* kidney, *Nature*, 287, 859, 1980.
31. Sokol, J. H., Lee, C. O., and Lupo, F. J., Measurement of the free calcium ion concentration in sheep cardiac Purkinje fibers with neutral carrier Ca²⁺-selective microelectrodes, *Biophys. J.*, 25, 143a, 1979.
32. Bers, D. M. and Ellis, D., Intracellular calcium and sodium activity in sheep heart Purkinje fibers, *Pfluegers Arch.*, 393, 171, 1982.
33. Murphy, G., Coll, K., Rich, T. L., and Williamson, J. R., Hormonal effects on calcium homeostasis in isolated hepatocytes, *J. Biol. Chem.*, 255, 6600, 1980.
34. Simons, T. J. V., Measurement of free Ca²⁺ in red blood cells, *J. Physiol.*, 318, 38, 1981.
35. Lew, V. L., Tsien, R. Y., Miner, C., and Bookchin, R. M., Physiological [Ca²⁺]ᵢ level and pump-leak turnover in intact red blood cells measured using an incorporated Ca chelator, *Nature*, 298, 478, 1982.
36. Lew, V. L., Tsien, R. Y., Miner, C., and Bookchin, R. M., Physiological [Ca²⁺]ᵢ level and pump-leak turnover in intact red blood cells measured using an incorporated Ca chelator, *Nature*, 298, 478, 1982.
37. Rink, T. J., Smith, S. W., and Tsien, R. Y., Cytoplasmic free Ca²⁺ in human platelets. Ca²⁺ thresholds and Ca-independent activation for shape-change and secretion, *FEBS Lett.*, 148, 21, 1982.
38. Litchman, A. H., Segel, G. B., and Litchman, M. A., An ultrasensitive method for the measurement of human leukocyte calcium: lymphocytes, *Clin. Chim. Acta*, 97, 107, 1979.
38a. Tsien, R. Y., Pozzan, T., and Rink, T. J., T-cell mitogens cause early changes in cytoplasmic free Ca²⁺ and membrane potential in lymphocytes, *Nature*, 295, 68, 1982.
39. Cobbold, P. H., Measurement of cytoplasmatic free calcium in an amoeba and in smaller cells using aequorin, in *Detection and Measurement of Free Ca²⁺ in Cells*, Ashley, C. C. and Campbell, A. K., Eds., Elsevier/North Holland, Amsterdam, 1979, 245.
40. Rink, T. J., Tsien, R. Y., and Warner, A. E., Free calcium in *Xenopus* embryon measured with ion-selective microelectrodes, *Nature*, 283, 658, 1980.
41. Cuthbertson, K. S. R., Whittingham, D. N., and Cobbold, P. H., Free Ca²⁺ increases in exponential fases during mouse oocyte activation, *Nature*, 294, 754, 1981.

Chapter 2

THE Ca^{2+} HOMEOSTASIS

A. F. Rega

I. HOW DOES Ca^{2+} ENTER AND LEAVE THE CELL?

Results in Table 2 show that the concentration of Ca^{2+} in the cytosol of cells at rest [Ca$^{2+}_i$] is on the order of 10^{-7} M. Since the concentration in the surrounding media [Ca$^{2+}_o$] is near 10^{-3} M, the influx of Ca$^{2+}_o$ is favored by a concentration ratio [Ca$^{2+}_o$]/[Ca$^{2+}_i$] close to 10^4. To this it has to be added the membrane potential, negative in the cell interior.

There are several mechanisms that have been proposed for the movement of Ca^{2+} across the plasma membrane of resting cells. Some are present in excitable as well as in nonexcitable cells and others have been described only in one of the two cell classes.

A. Excitable Cells

Present knowledge on the Ca^{2+} inflow and outflow across the plasma membrane of excitable cells comes from studies on nerve and muscle cells.

Plasma membranes from excitable cells are tight to Ca^{2+}. The permeability of the membrane of squid giant axons calculated as the ratio of the passive influx and extracellular concentration of Ca^{2+} [1] can be calculated to be about 10^{-8} cm·sec^{-1}, a lower value than those for Na$^+$ and for K$^+$ which have been reported to be 4×10^{-4} and 80×10^{-4} cm·sec^{-1}, respectively.[2] Membrane depolarization increases the passive movement of Ca^{2+} down the electrochemical gradient in nerve and muscle cells. The likely structure through which Ca^{2+} permeates are channel-forming proteins embedded in the lipid bilayer of the membrane. These channels are modulated by neurotransmitters and drugs and are controlled by voltage-dependent gating, that is, their opening and closing are the result of changes in membrane potential.[3] Therefore, excitable cells are endowed with voltage-sensitive channels for Ca^{2+} which allow a transient influx of Ca^{2+} during excitation; these channels are described in Chapter 3.

In the axon from squid nerve cells at rest, total Ca^{2+} influx can be divided into two components: (1) a Na^+_i-dependent component and (2) a Na^+_i-, Ca$^{2+}_i$-, and ATP-insensitive component.[4-6] The first component is a Ca$^{2+}_o$-Na^+_i exchange. Under intracellular conditions resembling the physiological one and for an external Ca^{2+} concentration similar to that of the sea water (10 mM), the rate of Ca^{2+} entrance in exchange with Na$^+$ is near 0.004 pmol/cm^2/sec, but it can increase up to 80 times in Li$^+$ sea water.[4] The stoichiometry of the exchange is 3 to 5 Na^+_i per Ca$^{2+}_o$ and the apparent affinity for Ca^{2+} is 50 to 100 mM. The rate of the exchange is drastically lowered by removal of ATP and increased by electrical depolarization. It is also inhibited by intracellular EGTA at concentrations that give physiological concentrations of Ca^{2+}, suggesting that the influx depends on Ca^{2+} at the cell interior. The component of Ca^{2+} influx that is insensitive to Na^+_i, Ca^{2+}, and ATP in squid axons is about 0.040 pmol/cm^2/sec at 4 mM Ca$^{2+}_o$ [5] and up to 100 mM Ca^{2+}, it increases almost linearly with the concentration of Ca$^{2+}_o$ as if it were a leak pathway.[4]

From an operational point of view, the Ca^{2+} efflux from squid axons at rest can be divided into three components: (1) a Na^+_o-dependent component; (2) a Ca$^{2+}_o$-dependent component, and (3) a Na^+_o- and Ca$^{2+}_o$-independent component.[4-6] The first component represents a carrier-mediated Ca$^{2+}_i$-Na^+_o exchange; the second a Ca^+_o-Ca$^{2+}_o$ exchange; and the third the active transport catalyzed by the Ca^{2+} pump.

Under physiological conditions, the efflux of Ca^{2+} via the Na^+-Ca^{2+} exchanger represents about 30 to 50% of the total calcium efflux. Its rate under physiological conditions is 0.03 to 0.05 pmol/cm²/sec, but it can reach values as high as 2 pmol/cm²/sec.[4] The stoichiometry is about $3Na^+$ per Ca^{2+}. Li^+, K^+, or $choline^+$ cannot replace Na^+ as the counter-ion. The apparent dissociation constant of the system for Ca^{2+}_i is 0.73 μM and for Na^+_o is 50 to 100 mM. The gradient of Na^+ created by the Na^+ pump across the plasma membrane is theoretically capable of providing the energy necessary for this efflux of calcium in exchange for Na^+_o, but ATP with a low affinity (Km near 300 μM) stimulates the movement of calcium by increasing the apparent affinities of the transport system for both Ca^{2+}_i and Na^+_o.[4] The mechanism of the ATP effect is still uncertain, but the idea of a regulatory role for ATP is favored. The Ca^{2+}_i-Ca^{2+}_o exchange is about 0.01 pmol/cm²/sec and under physiological conditions, represents less than 10% the total efflux of calcium.[4] The apparent affinity for intracellular Ca^{2+} is 1 mM and the stoichiometry is not well known.

The uncoupled efflux (also called "residual" Ca^{2+} efflux[4]) represents the active transport of Ca^{2+} that takes place through the Ca^{2+} pump of the plasma membrane. The term *uncoupled* has been used to indicate that in nerve cells, the extrusion of Ca^{2+} by this mechanism is apparently not accompanied by the translocation of either Ca^{2+} or Na^+ as the counter-ion. It is sensitive to vanadate, is independent of any ionic gradient, is near 0.04 pmol/cm²/sec and represents at least 40% of total calcium efflux under near-physiological conditions.[5] At nonlimiting concentrations of Ca^{2+}_i, it can reach values of 0.15 pmol/cm²/sec. It is described with more detail in Chapter 4.

B. Nonexcitable Cells

The passive permeability of nonexcitable cells to Ca^{2+} is thought to be very low, but it seems to rely upon the metabolic condition of the cell. No net calcium uptake can be detected in fresh human red blood cells suspended in Ca^{2+}-containing buffers at 37°C. In the cold, that is, under conditions in which active transport should be very low, the intracellular Ca^{2+} content of human red blood cells after 1 week is 1.8% of the extracellular Ca^{2+} at zero time, indicating that the passive permeability of the red blood cell membrane to Ca^{2+} is very low.[7] After the cells have been depleted of ATP, they gain Ca^{2+} at a rate up to 30 μmol/l cells/hr.[8,9] The passive efflux of calcium from rat liver microsomes preloaded with 10^{-4} M Ca^{2+} is 0.2 nmol Ca^{2+}/mg protein/min. This value is 2.5 to 5.0 times higher if the microsomes were made out of the plasma membrane of cells from ischemic livers which are depleted of metabolic energy.[10] These findings suggest that calcium permeability in nonexcitable cells increases after depletion of metabolic energy.

The information available on the mechanisms of Ca^{2+} permeation in nonexcitable cells comes from studies made on red blood cells. Red blood cells are advantageous for this sort of study, because they are devoid of intracellular organelles. Ferreira and Lew[8] found that the rate of Ca^{2+} influx in red blood cells that had been depleted of ATP increases with external Ca^{2+} along a curve that levels off at a rate of 30 μmol/l cells/hr and reaches its half-maximum value at about 1 mM external Ca^{2+}. The nonlinear response to external Ca^{2+} makes it reasonable to think that the net Ca^{2+} influx in red blood cells takes place throughout a limited number of sites in the membrane and hence that it behaves as if it were a facilitated diffusion process. On the other hand, the influx of radioactive Ca^{2+} is higher in calcium-containing cells than in calcium-free cells. This has been taken as an evidence that at least part of the calcium entry in red blood cells takes place through a Ca^{2+}-Ca^{2+} exchanger without net accumulation of calcium in the cytosol. No Na^+_i-Ca^{2+}_o exchange has been demonstrated to take place in human red blood cells.

The main pathway for the outflow of Ca^{2+} in nonexcitable cells is the active transport through the Ca^{2+} pump mechanism, whose rate can reach values much larger than that of Ca^{2+} transport driven by the electrochemical gradient. In ATP-depleted red blood cells, only the passive mechanisms operate. In these cells, calcium efflux could be divided into two components: one depends on calcium on the external surface of the membrane and represents a Ca^{2+}_i-Ca^{2+}_o exchange and the other is independent of external calcium.[8] Both components saturate as a function of internal calcium. The component that does not depend on external calcium shows a maximum efflux rate of 15 μmol Ca^{2+}/ℓ cells/hr and may respond to a facilitated diffusion mechanism. The Ca^{2+}_i-Ca^{2+}_o exchange has a maximum rate of about 70 μmol/ℓ cells/hr and is half-maximum at 30 μmol Ca^{2+}_i/ℓ cells at 15 μM Ca^{2+}_o when Ca^{2+} concentration at the opposite surface of the membrane is saturating.

Although there is no experimental evidence, it is simple to think that a single Ca^{2+}_i-Ca^{2+}_o exchange mechanism operates during passive Ca^{2+} inflow and passive Ca^{2+} outflow in red blood cells. This seems not to be the case for the nonexchange efflux and the nonexchange influx which, in ATP-depleted red blood cells, appear to be kinetically distinct.[8]

II. TRANSPORT OF Ca^{2+} BY INTRACELLULAR ORGANELLES

It seems that Ca^{2+} enters the mitochondria through a path which is different from that that Ca^{2+} uses to leave the organelle. Ca^{2+} is transported into respiring mitochondria by a secondary active transport process (as defined by Stein[11]) driven by the membrane potential gradient (negative inside) across the inner mitochondrial membrane generated either by the electron transport or by the hydrolysis of ATP.[12] Transported calcium carries two positive charges that are compensated by two H^+ extruded by the respiratory chain. Mitochondrial Ca^{2+} uptake driven by the membrane potential can be inhibited by the glycoprotein stain ruthenium red or La^{3+} acting directly on the calcium carrier and by the uncoupler of oxidative phosphorylation FCCP depressing the membrane potential generated by the respiratory chain mechanism. Carafoli et al.[13] reported results showing that the rate of Ca^{2+} uptake by the mitochondria as a function of Ca^{2+} follows saturation kinetics with K_{Ca} about 5 μM and maximum rate 30 nmol/mg protein/min. Nichols[14] reported experiments in which the rate of Ca^{2+} uptake appears to be limited by the rate at which the respiratory chain can expel protons, since on raising from 0.8 to 4.0 μM the concentration of Ca^{2+} in the suspending medium, the rate of calcium uptake by isolated rat liver mitochondria increases from 5 to 300 nmol/mg protein/hr. If the membrane potential that energizes the uptake of Ca^{2+} decreases sufficiently, the mechanism of Ca^{2+} uptake by the mitochondria becomes reversible and allows Ca^{2+} efflux to occur, a process that is inhibited by ruthenium red.[15] In the absence of membrane potential or in the presence of ruthenium red, a slow efflux of Ca^{2+} of about 1 to 5 nmol/mg protein/hr can be detected in mitochondria which have accumulated calcium. The rate of this Ca^{2+} efflux that takes place through a separate and independent pathway is almost constant once about 10 nmol Ca^{2+}/mg protein have accumulated in the mitochondria.[14] In heart, brain, and skeletal muscle mitochondria, the efflux of Ca^{2+} is activated by Na^+, and it has been suggested that it occurs at a Na_o-Ca^+_i exchange transporter. However, in liver, kidney, and smooth muscle mitochondria, the efflux of Ca^{2+} is not activated by Na^+, and it may represent a H^+_o-Ca^{2+}_i exchange.[16]

Bygrave[17] studied the properties of Ca^{2+} transport by a microsomal preparation derived from the endoplasmic reticulum of rat liver cells. There is a Ca^{2+} influx whose initial rate reaches values of 25 nmol/mg protein/hr. The uptake is an active process because it depends on MgATP with a Km of about 500 μM. The K_{Ca} is 1 μM or less and the uptake is little affected by ruthenium red. The time course of the influx can be described in terms of a rapid initial phase followed by a phase of much slower transport

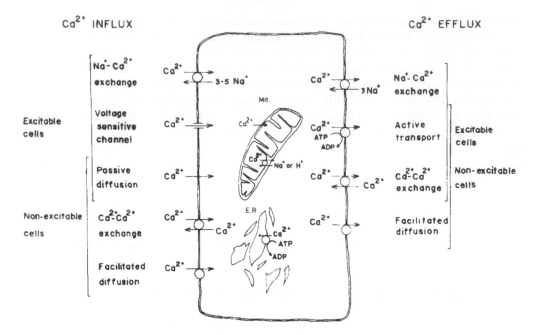

FIGURE 1.   The cellular mechanisms for Ca²⁺ influx and efflux. Mit: Mitochondrium, E.R.: Endoplasmic reticulum. (The scheme was drawn following G. J. Barrit's idea in Calcium transport across cell membranes: progress toward molecular mechanism, *TIBS*, 12, 322, 1981.)

which may be a consequence of a steady-state situation in which Ca²⁺ influx is either inhibited by internal Ca²⁺ or takes place together with Ca²⁺ efflux.

The sarcoplasmic reticulum membrane from muscle cells is one of the more active Ca²⁺ transporting systems. It couples the hydrolysis of ATP to the transport of Ca²⁺ from the cytoplasm to the sarcoplasmic reticulum with high affinity ($K_{Ca} < 10^{-6}$ M), and an initial rate of transport that in a microsomal preparation from white skeletal muscle was reported to be as high as 3600 nmol/mg protein/hr.[18]

Figure 1 shows the mechanisms and Table 1 summarizes and allows comparison of the kinetic parameters of the known pathways of Ca²⁺ transport across plasma membrane of different cells and intracellular organelles.

III. THE ROLE OF THE TRANSPORT MECHANISMS IN THE REGULATION OF Ca²⁺ CONCENTRATION IN THE CYTOSOL

It remains now to specify the role of each of the subcellular organelles and the transport systems in the regulation of Ca²⁺ concentration in the cytosol. Needless to say, the sarcoplasmic reticulum with its great capacity to store and release Ca²⁺ plays a pivotal role in the regulation of cytosolic Ca²⁺ in muscle cells. But the picture is more complicated in nonmuscle cells. The relative contribution of each transport system is difficult to evaluate, since knowing their specific activity and apparent affinity for Ca²⁺ tells nothing about their total capacity and relative contribution to the intracellular Ca²⁺ movements.

Nevertheless, present knowledge allows one to propose cycling of Ca²⁺ through independent influx and efflux pathways across the mitochondrial inner membrane as one of the various mechanisms that participate in the regulation of the concentration of Ca²⁺ in the cytosol. The magnitude of the Ca²⁺ fluxes across the mitochondrial membranes and their dependence on the concentration of Ca²⁺ on both sides of the membrane allow one to conclude that if mitochondria were exposed to Ca²⁺ concentrations

Table 1

KINETIC PARAMETERS OF Ca²⁺ TRANSPORT ACROSS DIFFERENT MEMBRANES IN RESTING CELLS

Membrane system	Mechanism	Rate under near physiological condition (nmol/mg protein/min)	Maximum rate (nmol/mg protein/min)	K_m (mM)	Comments	Ref.
			Ca²⁺ Influx			
Nerve[a] (Squid axon)	Ca²⁺ₒ-Na⁺ᵢ exchange	0.4		50—100 (Ca²⁺ₒ) 3—10 in Li⁺ sea water	Increased by ATP, Ca²⁺, Na⁺, and depolarization	4
	Passive diffusion	3.7 (at Ca²⁺ₒ 4 mM)			Increased by depolarization. Inhibited by La³⁺	5
Red blood cell	Facilitated diffusion		0.05	1 (Ca²⁺ₒ)	Activated by ATP depletion	8
	Ca²⁺ₒ-Ca²⁺ᵢ exchange		0.03		Activated by ATP depletion and by Ca²⁺ᵢ	8
Mitochondria	Secondary active transport	5	300 Limited by the rate of respiration	10⁻²	Inhibited by ruthenium red, Mg²⁺ La²⁺ and uncouplers	14
Endoplasmic reticulum	Active transport		25	10⁻³		17
Sarcoplasmic reticulum	Active transport		3,600	10⁻³	Inhibited by VO₃⁻	18
			Ca²⁺ Efflux			
Nerve[a] (Squid axon)	Ca²⁺ᵢ-Na⁺ₒ exchange	4.9		0.7 × 10⁻³ (Ca²⁺ᵢ)	Activated by ATP; inhibited by La²⁺ and depolarization	4
	Ca²⁺ᵢ-Ca²⁺ₒ exchange	0.98		1 (Ca²⁺ᵢ)		4
	Active transport	3.9	14.7	0.2 × 10⁻³	Inhibited by VO₃⁻	5—6

Table 1 (continued)
KINETIC PARAMETERS OF Ca²⁺ TRANSPORT ACROSS DIFFERENT MEMBRANES IN RESTING CELLS

Membrane System	Mechanism	Rate under near physiological condition (nmol/mg protein/min)	Maximum rate (nmol/mg protein/min)	K_{Ca} (mM)	Comments	Ref.
Red cell[b]	Facilitated diffusion		0.025			8
	Ca²⁺:Ca²⁺ exchange		0.12	3×10^{-2} (Ca²⁺₀)		8
	Active transport		66	10^{-3}	Inhibited by VO₃⁻	19
Mitochondria	Ca²⁺:Na⁺ or Ca²⁺:He⁺ exchange	1	5			14

[a]   Estimated assuming a membrane protein content of 0.613 µg/cm².
[b]   Estimated assuming a membrane protein content of 10 g/ℓ cells.

below 0.8 μM, the influx of Ca^{2+} would be less than the efflux, and a net loss of Ca^{2+} through the slow efflux pathway will occur until a steady state is reached. Furthermore, the magnitude and the steep response of the Ca^{2+} uptake in mitochondria to changes in Ca^{2+} concentrations between 1 and 4 μM commented upon before[14] suggest that mitochondria can participate in coarse regulation of cytosolic Ca^{2+}.

Becker et al.[20] made an original experimental approach to study the regulation of Ca^{2+} concentration by mitochondria and endoplasmic reticulum from rat liver cells. They made electrode measurements of Ca^{2+} concentration in a medium of composition similar to that of the cytosol. When respiring mitochondria were suspended in this artificial cytosol, the steady-state Ca^{2+} concentration in the medium was 0.5 μM. Upon addition of a microsomal fraction derived from the endoplasmic reticulum, there was a rapid drop in the steady-state level of Ca^{2+} to about 0.2 μM, followed by a slow release of about 50% of the mitochondrial Ca^{2+}. The concentration of Ca^{2+} in the medium remained at 0.2 μM after a suspension of hepatocytes that had been permeabilized with digitonin was added, suggesting that the conditions maintained by the mitochondria and the endoplasmic reticulum vesicles in the artificial cytosol approximate the conditions of the cytosol in vivo. These observations give experimental support to the idea that the role of respiring mitochondria is the coarse regulation by the rapid accumulation and slow release of large amounts of Ca^{2+} and that endoplasmic reticulum, in the presence of respiring mitochondria, participates in the fine regulation of cytosolic Ca^{2+} concentration. The participation of endoplasmic reticulum in the regulation of cytosolic Ca^{2+} has recently received support from the finding that inositol triphosphate may stimulate the release of Ca^{2+} by the endoplasmic reticulum.[21] Data in Table 1 shows that in eukaryotic cells, the rate of the influx of Ca^{2+} through the plasma membrane is low and represents only a fraction of the maximum rate of the efflux through the Ca^{2+} pump. Excitable cells which allow penetration of extracellular Ca^{2+} after stimulation, in addition to the Ca^{2+} pump, possess transporters that catalyze the rapid outflow of Ca^{2+} through the plasma membrane to restore the initial concentration of Ca^{2+} in the cytosol.

The maintenance of the calcium content of the cell at a steady level for long periods under conditions in which there is a small but permanent inward leak of Ca^{2+} implies that the total amount of Ca^{2+} available to the machinery that binds Ca^{2+} in the cell interior must be subjected to a sort of control at the plasma membrane. This role has been assigned to the Ca^{2+} pump from plasma membrane which pumps out, to a infinite reservoir, the extracellular Ca^{2+} that enters the cell. The role of the Ca^{2+} pump of plasma membrane is therefore fundamental in determining the concentration of Ca^{2+} in the cytosol and hence in the whole of the cell.

REFERENCES

1. Baker, P. F., and McNaughton, P. A., The influence of extracellular calcium binding on the calcium efflux from squid axons, *J. Physiol.*, 276, 127, 1978.
2. Rojas, E., Ion permeability, in *New Comprehensive Biochemistry-Membrane Transport*, Vol. 2, Bonting, S. L. and de Pont, J. J. H. H. M., Eds., Elsevier/North-Holland, Amsterdam, 1981, chap. 3.
3. Reuter, H., Calcium channel modulation by neurotransmitters, enzymes and drugs, *Nature*, 301, 569, 1983.
4. Baker, P. F., The regulation of intracellular calcium in giant axons of *Loligo* and *Myxicola*, *Ann. N.Y. Acad. Sci.*, 307, 250, 1978.
5. DiPolo, R. and Beaugé, L., Mechanisms of calcium transport in the axon of the squid and their physiological role, *Cell Calcium*, 1, 147, 1980.

6. DiPolo, R. and Beaugé, L., The calcium pump and sodium-calcium exchange in squid axons, *Ann. Rev. Physiol.*, 45, 313, 1983.

7. Schatzman, H. J. and Vincenzi, F. F., Calcium movements across the membrane of human red cells, *J. Physiol.*, 201, 369, 1969.

8. Ferreira, H. G. and Lew, V. L., Passive Ca transport and cytoplasmic Ca buffering in intact red cells, in *Membrane Transport in Red Cells*, Ellory, J. C. and Lew, V. L., Eds., Academic Press, London, 1977, 53.

9. Lew, V. L., Tsien, R. Y., Miner, C., and Bookchin, R. H., Physiological [Ca²⁺]ᵢ level and pump-leak turnover in intact red cells measured using an incorporated Ca chelator, *Nature*, 298, 478, 1982.

10. Chien, K. R., Abrams, J., Scroni, A., Martin, J. T., and Farber, J. L., Accelerated phospholipid degradation and associated membrane dysfunction in irreversible, ischemic liver cell injury, *J. Biol. Chem.*, 253, 4809, 1978.

11. Stein, W. D., Concepts of mediated transport, in *New Comprehensive Biochemistry Membrane Transport*, Vol. 2, Bonting, S. L. and de Pont, J. J. H. H. M., Eds., Elsevier/North-Holland, Amsterdam, 1981, chap. 5.

12. Racker, E., Fluxes of Ca²⁺ and concepts, *Fed. Proc. Fed. Am. Soc. Exp. Biol.*, 39, 2422, 1980.

13. Carafoli, E., Crompton, M., Malmstrom, K., Sigel, E., Salzmann, M., Cheisi, M., and Affolter, H., Mitochondrial calcium transport and the intracellular calcium homeostasis, in *Biochemistry of Membrane Transport*, Semenza, G. and Carafoli, E., Eds., Springer-Verlag, Berlin, 1977, 535.

14. Nichols, D. G., The regulation of extramitochondrial free calcium ion concentration by rat liver mitochondria, *Biochem. J.*, 176, 463, 1978.

15. Nichols, D. G. and Scott, I. D., The regulation of brain mitochondrial calcium-ion transport, *Biochem. J.*, 186, 833, 1980.

16. Nichols, D. G., Some recent advances in mitochondrial calcium transport, *TIBS*, 6, 36, 1981.

17. Bygrave, F. L., Properties of energy-dependent calcium transport by rat liver microsomal fraction as revealed by initial-rate measurements, *Biochem. J.*, 170, 87, 1978.

18. Inesi, G. and Scarpa, A., Fast kinetics of adenosine triphosphate dependent Ca²⁺ uptake by fragmented sarcoplasmic reticulum, *Biochemistry*, 11, 356, 1972.

19. Kratje, R. B., Garrahan, P. J., and Rega, A. F., The effects of alkali metal ions on active Ca²⁺ transport in reconstituted ghosts from human red cells, *Biochim. Biophys. Acta*, 731, 40, 1983.

20. Becker, G. L., Fiskum, G., and Lehninger, A. L., Regulation of free Ca²⁺ by liver mitochondria and endoplasmic reticulum, *J. Biol. Chem.*, 255, 9009, 1981.

21. Sombyo, A. P., Cellular site of calcium regulation, *Nature*, 309, 516, 1984.

Chapter 3

CALCIUM AND CELL FUNCTION

P. J. Garrahan

I. INTRODUCTION

The crucial role of Ca^{2+} in the regulation of cell functions has been appreciated for a long time. However, in the last decade this field has undergone an explosive growth. This was so mainly because of the enormous advance in the knowledge of the biochemistry of the effects of Ca^{2+} that was made possible by the discovery of the widely distributed intracellular Ca^{2+}-binding proteins that mediate most of the effects of Ca^{2+}. The study of the cellular effects of Ca^{2+} has become a central subject in physiology, biochemistry, and cell biology.

It is far beyond the scope of this book to provide a comprehensive review of the effects of Ca^{2+} on cell function or of the huge amount of literature that has appeared during the last years on very different aspects of this subject. We will therefore circumscribe our analysis to some well-characterized examples of Ca^{2+}-mediated cell functions. We will cite some key papers and reviews to which the interested reader is referred for further information.

II. THE MESSENGER ROLE OF Ca^{2+}

It is now well established that cells increase their cytosolic levels of Ca^{2+} in response to many stimuli and that that increase is the primary event that triggers the cellular response. No case is known in which the response to a stimulus leads to a decrease in the cytosolic concentration of Ca^{2+} (see Reference 1). Cytosolic Ca^{2+} therefore acts as a "second messenger" that delivers to the effector system the signal of the stimulus. In eukaryotic cells, the role of Ca^{2+} as a second messenger is as important as the more widely known messenger role of cAMP. The nucleotide and Ca^{2+} may mediate responses to different stimuli, but frequently the responses to both are coordinated, and in some cases, cAMP and Ca^{2+} affect in different ways the same effector system. The suitability of Ca^{2+} chemistry for the role of Ca^{2+} as a second messenger as compared with the more abundant monovalent cations has been examined by Urry and co-workers[2,3] who postulate that it derives from the greater restrictions for the movements of Ca^{3+} across the cell membrane and its greater affinity for binding to proteins and polypeptides. The main experimental support for the role of Ca^{2+} as a second messenger comes from studies that: (1) demonstrate a correlation between the changes in the cytosolic Ca^{2+} and the time course of a stimulus, (2) show that a cellular response can be reproduced in the absence of its physiological stimulus by artificially raising the cytosolic level of Ca^{2+}, and (3) show a direct dependence on Ca^{2+} of an enzyme or other intracellular system.

After the stimulus has ceased, the concentration of Ca^{2+} is returned to its resting level by means of the same mechanisms that are responsible for the maintenance of the resting steady-state level of cytosolic Ca^{2+} and which are discussed in Chapter 2.

A. The Mechanism of the Increase of Cytosolic Ca^{2+} in Stimulated Cells

Depending on the cell and on the stimulus, cytosolic Ca^{2+} concentration may increase, either as consequence of the inflow of extracellular Ca^{2+} or because Ca^{2+} is released from intracellular stores (see Reference 4). In the first case, the response to

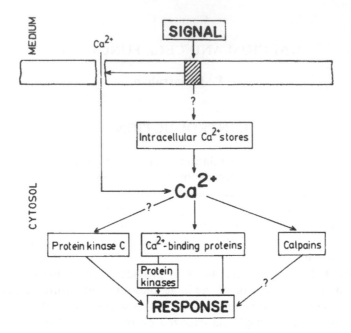

FIGURE 1.   A schematic representation of the messenger role of cytosolic Ca^{2+}. A signal acting on the plasma membrane, usually at a specific receptor, may increase the concentration of cytosolic Ca^{2+}, either because it promotes the opening of Ca^{2+} channels in the plasma membrane or because it induces the release of Ca^{2+} stored in cell organelles. Little is known about the mechanisms by which the signal reaches the intracellular organelles to promote the release of Ca^{2+}. When cytosolic Ca^{2+} concentration rises, specific sites in Ca^{2+}-binding proteins become occupied. The complex between Ca^{2+} and the Ca^{2+}-binding protein may trigger the response either directly or through the activation of a protein kinase which catalyzes the phosphorylation of the system in charge of the response. Protein kinase C may also participate. However, it is not yet clear whether elevated Ca^{2+} in addition to diacylglycerol (see text) is needed for the physiological activation of this enzyme. Calpains are Ca^{2+}-dependent proteases. Their role in Ca^{2+}-mediated cell responses is not yet established.

the stimulus disappears when the cells are suspended in a Ca^{2+}-free medium, whereas in the second case, the stimulus may be effective for long periods of time even in the complete absence of extracellular Ca^{2+}. It seems likely that release from intracellular stores is the preferred mechanism in large cells and when the whole cell must respond simultaneously. An example of this is the contraction of skeletal muscle in which, if the source of Ca^{2+} were that flowing in from the extracellular medium, diffusion delays would hamper the synchronous response of all myofibrils.

Present experimental evidence seems to favor the idea that in most cells the main intracellular store of messenger Ca^{2+} is the endoplasmic reticulum rather than the mitochondria (for references, see Reference 5). This is in keeping with the fact that the cytosolic Ca^{2+} concentration in stimulated cells is usually below that needed to switch on high-capacity and relatively low-affinity mitochondrial Ca^{2+} transport system.

1. Ca^{2+} Channels in Excitable Membranes

Regardless of the source, the flow of Ca^{2+} into the cytosol that follows the stimulus is a net flow down an electrochemical potential gradient. In most cases, this probably is the consequence of a transient increase in the permeability to Ca^{2+} of the plasma membrane or of the membrane that surrounds the intracellular Ca^{2+} store.

Electrophysiological studies have shown unequivocally that the plasma membrane of cells that respond to stimuli with changes in membrane potential possess channels that selectively allow the entry of Ca^{2+} down its electrochemical potential gradient (for a review and references, see Reference 6). These channels have two main functions: to allow the participation of Ca^{2+} currents in the rising phase of the action potential; and to allow the inflow of extracellular Ca^{2+} that leads to the rise in cytosolic Ca^{2+} concentration and the consequent cellular response to the stimulus.

Ca^{2+} channels are controlled by voltage-dependent gating, that is, their opening or closing kinetics are a consequence of changes in membrane potential. In resting cells, the channels are closed. The probability that a channel will open increases steeply with depolarization. The channels show both time- and voltage-dependent inactivation. Usually they open at more positive membrane potentials than the Na^+ channels which participate in the rising phase of the action potential. The Ca^{2+} channels are one of the most selective of the several kinds of channels which can be detected in excitable membranes. For example, the selectivity for Na^+ of the Na^+ channels is at least one order of magnitude less than the selectivity for Ca^{2+} of the Ca^{2+} channels.

Although Ca^{2+} channels are primarily regulated by the membrane potential, their properties are modulated by neurotransmitters, hormones, and drugs. In some cases, modulation seems to imply phosphorylation of the channel or of membrane proteins closely associated with it.

Single-channel conductances have been measured in several preparations. This allows one to calculate the surface density of channels which varies in different tissues and in different regions of the same cell. In some cases, the variations have a clear physiological meaning. In skeletal muscle, for instance, they are more abundant in the T tubules which is the region where the signal of the action potential is transmitted to the sarcoplasmic reticulum through a mechanism that is not yet understood. In squid neurons, the channel density is small in the axon and large in the synaptic region, a fact which is in keeping with the lack of requirement of Ca^{2+} for the generation of the action potential and the absolute requirement of this cation for the release of neurotransmitter at the synapse.

Ca^{2+} channels can be blocked to various degrees and with different specificities by inorganic ions such as Mg^{2+}, Ni^{2+}, Co^{2+}, Cd^{2+}, and La^{3+}. Some natural toxins like veratridine, grayanotoxin, and batrachotoxin are also blockers of the Ca^{2+} channels. In addition, several classes of drugs are more or less selective blockers of Ca^{2+} channels, and their use has contributed to the understanding of some properties of these channels. These drugs include the phenylalkylamines, the benzodiazepines, and the dihydropyridines. The dihydropyridines are the most potent blockers.[7-10] Specific binding sites for these compounds have been found in plasma membranes from many excitable cells. Binding seems to be Ca^{2+}-dependent[7] and takes place with high affinity (Kd = 0.05 to 1 nM). The sites for these compounds seem to be different than the binding sites of other Ca^{2+}-channel antagonists such as the inorganic ions and natural toxins.[7,8] Dihydropyridine binding sites have been solubilized from brain membranes with digitonin. They have been tentatively identified as large membrane glycoproteins which retain the binding sites for the other Ca^{2+}-channel antagonists.[8] Studies on the radiation inactivation of the specific binding of one of the dihydroxynitropyridine (nitrendipine) indicates that in both brain and muscle cells the binding site belongs to a large protein (M_r around 230,000). It is interesting to note that the apparent size of the Ca^{2+} channel is similar to that measured both by radiation inactivation and by hydrodynamic methods for the tetrodoxin-sensitive Na channel, suggesting that there may be structural similarities between these two classes of voltage-dependent ionic channels (see Reference 9). It has been recently shown that some dihydropyridines have an opposite effect to that mentioned above in the sense that they induce the opening of the Ca^{2+} channels.[10]

2. Cell Membrane Phospho- and Polyphosphoinositides and Receptor-Mediated Ca² ⁺ Mobilization

A variety of biologically active substances acting at receptors on the plasma membrane provoke the increase in the turnover of membrane phosphoinositides (PI) and polyphosphoinositides. PIs have one phosphate group in the inositol ring, whereas polyphosphoinositides have one or two additional phosphate groups. Polyphosphoinositides include phosphatidylinositol 4-phosphate (PIP) and phosphatidylinositol 4,5-bis-phosphate (PIP2) (see Reference 5). The increase in the turnover of these compounds is associated with the mobilization of intracellular Ca² ⁺ and not with the synthesis of cAMP. Stimulation of turnover does not require extracellular Ca² ⁺ and is unaffected by changes in intracellular Ca² ⁺ concentration (see Reference 5).

Receptors involved in this type of response include the muscarinic receptor for acetylcholine, the α1 adrenergic receptor, the thrombin receptor of platelets, and one type of receptor for vasopressin.

In 1975, Michell[11] proposed that after binding to the receptors of ligands that promote the PI turnover, the hydrolysis of PI by a phospholipase is activated with the consequent release of diacylglycerol and this event would lead to the opening of a channel which would allow the inflow of extracellular Ca² ⁺ into the cytosol. A plasma membrane kinase would convert the diacylglycerol into phosphatidate which would then participate in the resynthesis of PI in the smooth endoplasmic reticulum.[12] Michell's scheme has led to controversy (see, for instance, References 13 and 14).

A novel possibility for the interpretation of the response of membrane PI, PIP, and PIP2 to chemical signals is provided by the recent findings indicating that some products of the hydrolysis of these compounds (namely diacylglycerol and inositol triphosphate) act as intracellular messengers. In fact, it is now known that: (1) diacylglycerol is essential for activation of phosphokinase C, an enzyme which requires Ca² ⁺; (The properties of this enzyme and its possible role in mediating signals delivered by Ca² ⁺ will be discussed in another section.) and (2) in exocrine pancrease[15,16] and in hepatocytes,[17] physiological concentrations of inositol-1,4,5-triphosphate release Ca² ⁺ from nonmitochondrial intracellular stores. The intracellular organelle affected by this compound probably is the endoplasmic reticulum.

3. The Release of Ca² ⁺ from Intracellular Stores

The best-studied example of this phenomenon is that of skeletal muscle. In this system, it is well known that contraction is triggered by the massive release of Ca² ⁺ from the sarcoplasmic reticulum that follows the propagation of the action potential along the T tubules. The connection between the electric signal and the release of Ca² ⁺ is not yet clear, and both electrical and/or chemical signals have been implied in this process (see References 4 and 5).

In other types of cells such as the hepatocyte, α1 adrenergic agonists, glucagon, vasopressin, and angiotensin act at receptors in the plasma membrane and release Ca² ⁺ from an intracellular store. Mitochondria have been proposed a few years ago as the likely candidates for this role.[18] However, as we have already mentioned, more recent experimental work (see Reference 5) seems to indicate that Ca² ⁺ accumulated by the endoplasmic reticulum is the main intracellular store of Ca² ⁺ released upon stimulation.

It now seems likely that the signals acting at the cell membrane are transmitted to the intracellular Ca² ⁺ stores by means of inositol triphosphate.[5,17] If this proves to be true, inositol triphosphate would be the second messenger and cytosolic Ca² ⁺ the third messenger of the signal delivered by stimuli that act through this mechanism.

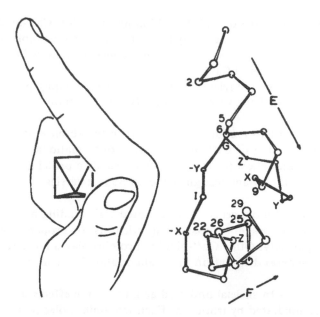

FIGURE 2. The homolog Ca²⁺ binding domain in Ca²⁺-bind-
ing proteins. Shown is the crystal structure of the Ca²⁺-binding
domain of parvalbumin. This consists of a segment of α-helix
(residues 9 through 11), a loop containing the calcium ligands
(10 to 21), and a second segment of α-helix (20 to 30). The
steric relationships of the helix-loop-helix is approximated by
the extended forefinger, clenched middle finger, and thumb of
a right hand. (From Kretsinger, R. H., *Neurosci. Res. Program
Bull.*, 19, 226, 1981. With permission.)

III. Ca²⁺-BINDING PROTEINS

A. General Properties

In most cases where Ca²⁺ acts as the second messenger of a stimulus, the chemical
signal that initiates the response is not Ca²⁺ itself, but the complex between Ca²⁺ and a
member of a special class of proteins named Ca²⁺-binding proteins. These proteins are
found in the cytosol or on a membrane facing the cytosol and are able to bind Ca²⁺
when cytosolic Ca²⁺ concentration rises above its physiological resting level.

The general properties of Ca²⁺-binding proteins have been clearly stated by
Kretsinger[19,20] and can be summarized as follows:

1. Ca²⁺-binding proteins show a high degree of structural homology, suggesting that
    they are derived from a common ancestor.
2. On the basis of the crystal structure of parvalbumin,[21] it is likely that their Ca²⁺-
    binding domains are formed by two alpha helix segments, about 10 amino acid
    residues long, separated by a beta turn, also about 10 amino acids long, which
    surrounds the bound Ca²⁺ (Kretsinger's EF-hand) (Figure 2). Depending on the
    protein, one to four EF-hands are present per molecule. The Ca²⁺-binding sites
    in the EF-hands use a pattern of three to five carboxylate groups and three to one
    neutral oxygen ligands.
3. As a consequence of the above-mentioned structure of the Ca²⁺ binding site, the
    affinity for Ca²⁺ of the Ca²⁺-binding proteins is very high, and its value is such
    that the sites for Ca²⁺ will be occupied in stimulated cells and free in resting cells.
    This has obvious physiological advantages.

4. The chemical properties of their Ca^{2+}-binding sites allow Ca^{2+}-binding proteins to have a high Ca^{2+} to Mg^{2+} selectivity ratio. This makes it possible for them to bind Ca^{2+} preferentially despite the intracellular concentration of Mg^{2+}, which is 1000 times higher than that of Ca^{2+}.

5. The rate of binding and release of Ca^{2+} from Ca^{2+}-binding proteins is fast. This is needed in view of the very quick on-off rates of many Ca^{2+}-mediated cell responses.

6. Ca^{2+}-binding proteins are functionally inert in the absence of bound Ca^{2+}. In all of them, the binding of Ca^{2+} induces large conformational changes. It is likely that these changes allow the target system, usually a protein, to interact with the Ca^{2+}-protein complex and recognize the message it carries.

The best-known and more generally distributed Ca^{2+}-binding protein is calmodulin, which mediates Ca^{2+} messages in most eukaryotic cells. Its properties will be described in some detail separately. Other Ca^{2+}-binding proteins play a more restricted role and have a much less general distribution than calmodulin. These include:

- Troponin C[21] — In skeletal and cardiac muscle, the effects of Ca^{2+} ions on contraction are mediated by troponin. Each troponin molecule is a trimer of three dissimilar units. One of these is troponin C, a small protein (M_r 18,000) with four Ca^{2+}-binding sites of which probably two participate in a contraction-relaxation cycle of contractile proteins. Binding of Ca^{2+} to troponin C induces a conformational change which allows the myosin-binding site in G actin (the F actin monomer) to become exposed and thus combine with the energized myosin head, initiating in this way the molecular events of contraction. Troponin C has a high degree of homology with calmodulin and substitutes it with varying effectiveness in several biochemical reactions. In contrast with calmodulin, troponin C shows significant sequence variations in primary structure, even among different tissues of the same species. As will be discussed in detail later, it seems that the Ca^{2+}-troponin C complex is a physiological activator of phosphorylase kinase providing a mechanism for the synchronization of muscle contraction and energy supply.

- Parvalbumin[19,20] — This is a small (M_r 12,000) acidic protein found in high concentration in the sarcoplasm of skeletal muscle and whose physiological role is still not clear.

- Vitamin D-dependent Ca^{2+}-binding protein — This protein is present in highest concentration in intestine, kidney, and shell gland, tissues across which Ca^{2+} is transported in large amounts, but it is also found in smaller amounts in several other tissues. The protein is synthesized *de novo* by the effect of 1,25-dihydroxycolecalciferol, the hormonal metabolite of vitamin D3. The protein is small (M_r 28,000), acidic (pK near 4), and heat stable. It has four binding sites with high affinity for Ca^{2+} and low affinity for Mg^{2+}.[22-24]

Although the mechanisms discussed in the preceding paragraphs seem to be general and well established, recent experimental evidence suggest that in some cases, the effects of increases in cytosolic Ca^{2+} concentration are caused by the direct activation of an enzyme by Ca^{2+} without the mediation of a Ca^{2+}-binding protein. This might be the case of the enzymes phosphokinase C and calpain whose properties will be discussed later.

B. Calmodulin

In 1970, Cheung[25] described an activator of brain cyclic nucleotide phosphodiesterase, and Kakiuchi and Yamazaki[26] reported a factor specific for brain Ca^{2+}-dependent phosphodiesterase. Subsequently, Teo and Wang[27] identified the two factors as a Ca^{2+}-binding protein. The name "calmodulin" to designate the activator was proposed in 1978 by Cheung et al.[28] (For reviews and references on the general properties and distribution of calmodulin, see References 29 to 34).

1. Distribution

Calmodulin has been found in so many tissues of so many different organisms as to make it very likely that it is a component of all eukaryotic cells. The concentration of calmodulin in most tissues varies between 2 to 40 μmol/kg wet weight (see Reference 32). This is well above that necessary for the establishment of maximally effective concentrations of Ca^{2+}-calmodulin complex. Calmodulin concentration, therefore, does not seem to be limiting for the reactions in which it participates. The electroplax of electric eel, mammalian brain, and mammalian testis have a particulary high content of calmodulin. In tissue homogenates, calmodulin is distributed both in the soluble and particulate fractions. The partition between these fractions depends on the concentration of Ca^{2+} which increases the amount of the enzyme bound to the particulate fraction.

2. Chemical and Physical Properties

The chemical and physical properties of calmodulin of all sources examined from coelenterates to man are practically identical, indicating that the molecule has a high degree of evolutionary stability and justifying its lack of species and tissue specificity.[29,30] Calmodulin is a small acidic protein (isoelectric point around 4), composed of a single polypeptide chain of 148 amino acid residues with a high content of alpha helix (Figure 3). Calmodulin is extremely resistant to heat, acid, and other treatments that promote protein denaturation. Its amino acid composition is conspicuous due to the presence of 27 glutamate and 23 aspartate residues and the unusual amino acid ε-trimethyllysine. Tryptophan and cysteine are absent. The lack of tryptophan and the high phenylalanine/tyrosine ratio give calmodulin a characteristic UV absorption spectrum. The absence of cysteine and hydroxyproline is probably advantageous, since it allows calmodulin to assume a highly flexible tertiary structure, a feature which may be related to its ability to interact with very different target proteins.[30] The first complete amino acid sequence of calmodulin was reported by Watterson et al.[35] in 1980. Since then, amino acid sequences have been determined for calmodulins isolated from various sources. The results obtained so far indicate that at most, 8% of the amino acid residues vary throughout evolution (for references, see Reference 36).

3. Binding of Ca^{2+}

The tertiary structure of calmodulin seems to consist of four roughly similar domains, each of which contains a Ca^{2+}-binding site with the properties of Kretzinger's EF-hand. These sites are labeled I, II, III, and IV from the N to the C terminus of the polypeptide chain in Figure 3. The dissociation constants for Ca^{2+} from these sites is not the same in each of the four sites, although it is sufficient to saturate calmodulin at the cytosolic Ca^{2+} levels of stimulated cells. Mg^{2+} competes with Ca^{2+} at these sites. The identification of low- and high-affinity sites and the possible existence of cooperative and anticooperative interactions between the sites have been the subject of many recent studies,[36-38] and they will not be dealt with here. What seems likely is that Ca^{2+} binds to its sites in calmodulin in a step-wise fashion.[32,37] The stoichiometry of the active Ca^{2+}-calmodulin complex has not yet been determined and may vary in different enzymes. On the other hand, the affinity of calmodulin for Ca^{2+} may vary depending on whether calmodulin is bound or not to the target protein. These factors might provide additional "fine tuning" of the interactions of Ca^{2+}-calmodulin and the target system.

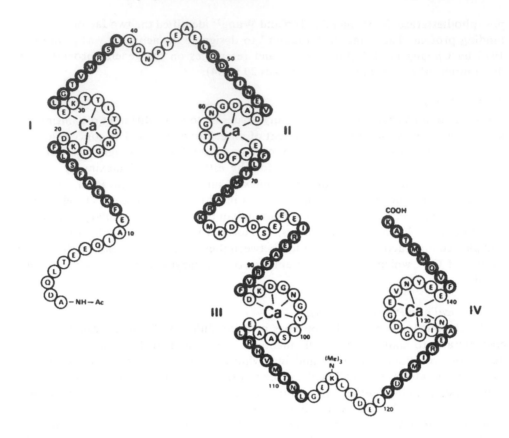

FIGURE 3.   The amino acid sequence of bovine brain calmodulin. Each of the four Ca²⁺-binding domains is flanked by two stretches of a α-helix (darker circles). (From Klee, C. B., Crowch, T. H., and Richman, P. J., *Ann. Rev. Biochem.*, 49, 489, 1980. With permission.)

Upon binding of Ca²⁺, calmodulin undergoes a large conformational change accompanied by a 5 to 10% increase in alpha helix content.[30] The Ca²⁺-dependent conformational change explains why Ca²⁺ converts the inactive conformer of calmodulin into an active one able to interact with the enzyme or protein that initiates the biological response to the stimulus that raised cytosolic Ca²⁺ concentration. Conformational changes have been detected by several means, including spectroscopic techniques, changes in reactivity towards chemical reagents, and altered susceptibility to enzymatic cleavage.[30,36,37] The Ca²⁺-mediated conformational changes take place in several (at least two) discrete steps, a fact which agrees with the proposed sequential nature of the addition of Ca²⁺ to calmodulin. Of particular interest seem to be the studies of LaPorte at al.[39] who showed that in the presence of Ca²⁺, calmodulin becomes able to bind hydrophobic reagents (Figure 4) and that the binding of these ligands antagonizes the interaction of Ca²⁺-calmodulin and effector proteins. The authors suggest that the conformational change that follows the binding of Ca²⁺ makes calmodulin expose a hydrophobic domain which is the interface for interaction with other proteins.

A number of drugs are able to block the activating effect of the Ca²⁺-calmodulin complex. These will be discussed in detail in Chapter 11. Most of these drugs also have other effects on cell function so that care must be exerted in identifying a calmodulin-dependent effect solely on the basis of its inhibition by a calmodulin-blocking agent. Of these, the most widely used are the phenothiazines, which have been shown to bind to calmodulin with high affinity in a Ca²⁺-dependent manner, yielding a complex unable to interact with target enzymes. The properties of the phenothiazines have been

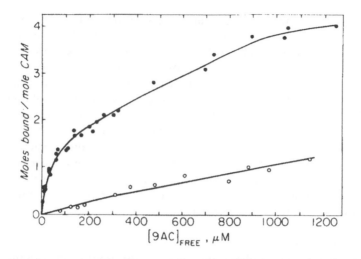

FIGURE 4. Equilibrium binding of the cationic amphiphile 9-anthoylcholine to calmodulin in the presence (O) and absence of 0.1 mM Ca^{2+}. (From LaPorte, D. C., Wierman, B. M. and Storm, D. R., *Biochemistry*, 19, 3814, 1980. With permission.)

used to purify calmodulin by affinity chromatography in phenothiazine-Sepharose[40,41] and to identify Ca^{2+}-calmodulin-dependent effects on the basis of their inhibition by these agents.

As will be shown in more detail in other sections of this chapter, the Ca^{2+}-calmodulin complex can act in two ways: one is directly on an effector system and the other is indirectly on a regulatory system, usually a protein kinase which through phosphorylation promotes or inhibits the activity of another enzyme. These two modes of action allow both fast and slow responses to be mediated by calmodulin and Ca^{2+}, a fact that contrasts with the other well-studied messenger cAMP, which always acts through the activation of protein kinases. Moreover, cAMP always acts through a single protein kinase which phosphorylates multiple substrates, whereas several Ca^{2+}-calmodulin-dependent kinases of more limited substrate specificity have been found.

IV. ENZYMES THAT DEPEND ON Ca^{2+} AND CALMODULIN

A. Enzymes Involved in Glycogen Metabolism

1. Skeletal Muscle Phosphorylase Kinase[42,43]

The main role of this enzyme is to catalyze the phosphorylation by ATP of serine residues of glycogen phosphorylase. This changes the enzyme from an inactive into an active form. Phosphorylase kinase also phosphorylates one of the serine residues of glycogen synthetase, but phosphorylation in this case results in a decrease in enzymic activity. Thus, as the result of the action of phosphorylase kinase, the rates through the opposing pathways of glycogenolysis and glycogen synthesis are changed in such a way that there is a net release of glucose from glycogen. In skeletal muscle, therefore, the release of Ca^{2+} from sarcoplasmic reticulum not only activates contraction, but also accelerates glycogenolysis for producing the ATP necessary to sustain muscle contraction. Phosphorylase kinase is also activated by cAMP-dependent protein kinases, but both the phosphorylated and dephosphorylated forms of the enzyme are totally dependent on Ca^{2+}. The enzyme is very large (M_r about 1,300,000) and has the subunit structure (α, β γ, δ)4. The α and β subunits are the components phosphorylated by cAMP-dependent protein kinases. γ is the catalytic subunit and δ is identical to calmodulin. Phosphorylation of the enzyme by cAMP-dependent protein kinases activates about 15-fold the enzyme at saturating concentrations of Ca^{2+} and increases substantially the affinity of the γ subunit for Ca^{2+} ions, allowing activation to occur when fewer Ca^{2+} ions are present. In the presence of Ca^{2+}, the dephosphorylated, but not the phosphorylated form of the kinase is strongly activated by a second calmodulin molecule (called the δ subunit). This effect is mimicked by troponin C, and there is evidence that in muscle, troponin C rather than calmodulin is the physiological activator. Compared to muscle phosphorylase kinase, considerably less is known of the liver enzyme. The liver enzyme is activated by cAMP, and both activated and nonactivated forms of the enzyme require and bind Ca^{2+} with high affinity. Since the γ1 adrenergic effects on glycogenolysis in liver seem to be mediated by the increase in cytosolic Ca^{2+}, it would seem reasonable to suppose that the behavior of the liver enzyme is not very different from that reported for the muscle enzyme.[16]

2. Glycogen Synthetase Kinase

Glycogen synthetase is phosphorylated on seven different serine residues by five different protein kinases. The effect of phosphorylation is to decrease the activity of the enzyme. A calmodulin-dependent glycogen synthetase kinase distinct from phosphorylase kinase was found in rat liver and phosphorylates glycogen synthetase at the so-called site 2.[44] A similar enzyme has been purified from rabbit skeletal muscle.[45] In contrast with the liver enzyme that seems to be an hexamer, the calmodulin-dependent phosphorylase is a dodecamer of M_r 696,000. Calmodulin-dependent glycogen synthetase kinase is not specific for glycogen synthetase, but is capable of phosphorylating many proteins in vitro including myosin light-chain kinase, histones, and acetyl-CoA carboxylase, which suggest that the enzyme may act on may substrates in vivo.

Glycogen synthetase kinase and a brain calmodulin-dependent protein kinase have been shown to have very similar properties. This has led to the suggestion that these kinases represent different isozymes of a multifunctional calmodulin-dependent protein kinase that mediates many of the actions of Ca^{2+} in many tissues. The name ''calmodulin-dependent multiprotein kinase'' has been proposed for this broad-specificity enzyme.[46]

3. Protein-Phosphatases

Four different protein-phosphatases (designated 1, 2A, 2B, and 2C), have been identified as capable of dephosphorylating the enzymes of glycogen metabolism.[47] Protein-phosphatase 2B is stimulated 10- to 20-fold by Ca^{2+}-calmodulin.[42,43,47] This enzyme is present at high concentration in skeletal muscle (where it has been purified to homogeneity[48]) and brain, but its concentration is low in liver. Because of its dependence on Ca^{2+} and calmodulin, in muscle it may be active only during contraction. This would provide a mechanism for Ca^{2+} to increase the rate of glycogen synthesis and to link the rate of synthesis to the strength of the previous contraction. Calcineurin is one of the major calmodulin-binding proteins of the brain (for references, see Reference 48). It seems to be very similar in structure, but it may be not identical to protein-phosphatase 2B. The enzyme is particularly rich in the neostriatum where it is localized mainly at postsynaptic sites within neuronal somata and dendrites (for references, see Reference 49). The enzyme is an heterodimer with a 61,000 M_r calmodulin-binding subunit (calcineurin A), and a 19,000 M_r subunit which binds four Ca^{2+} with high affinity (calcineurin B). It has been suggested that calcineurin A is the catalytic subunit subject to regulation by the Ca^{2+}-binding proteins calmodulin and calcineurin B acting at two different sites (see References 43 and 48). Tryptic proteolysis of the large subunit results in the loss of calmodulin binding and in stimulation of phosphatase activity. This suggests that calcineurin A consists of a catalytic domain resistant to proteolysis and a regulatory domain which exerts an inhibitory effect on catalysis, the inhibition being relieved by the binding of calmodulin to or by the proteolytic degradation of the regulatory domain.[50] During development, protein-phosphatase activity in the brain follows a time course that parallels synaptogenesis, suggesting a role for this enzyme in synaptic function.[49]

B. Enzymes Involved in Cyclic Nucleotide Metabolism

The Ca^{2+}-calmodulin complex activates adenylate cyclase, which initiates the cAMP signal, and phosphodiesterase, which switches off the signal by hydrolyzing cAMP. Only one of the several forms of cyclic nucleotide phosphodiesterase is activated by Ca^{2+}, namely the soluble enzyme that hydrolyzes cAMP with high Km (100 μM) and cGMP with low Km (5 to 20 μM). In contrast with the cytosolic distribution of the esterase, the cyclase is bound to the plasma membrane. Hence inflowing Ca^{2+} would first come into contact with the cyclase. On the other hand, activation by Ca^{2+} of the cyclase only takes place at submicromolar Ca^{2+} concentrations. At higher concentrations, Ca^{2+} inhibits the cyclase, but still activates the esterase. Therefore both the topography and the affinity of the enzymes involved in cAMP regulation allow a sequential activation of the enzymes.[26-28,32]

C. Enzymes Involved in Regulation of Muscle Contraction and Other Motile Processes
1. Myosin Light-Chain Kinase

In the presence of Ca^{2+} and calmodulin, this enzyme phosphorylates a single residue in the regulatory light-chain of myosin.[51,52] This is an essential event for the assembly of myosin into filaments, for actin-activated myosin-ATPase activity, and hence for contraction in smooth muscle and in nonmuscle tissues. In these systems, relaxation requires dephosphorylation of myosin light-chain by a specific phosphatase and the removal of Ca^{2+}. A phosphatase that dephosphorylates myosin light-chain has been recently purified from turkey gizzard muscle.[53] The enzyme is an heterotrimer of M_r 165,000. In the absence of Ca^{2+} and calmodulin, the enzyme is phosphorylated at two sites by a cAMP-dependent protein kinase. The dephosphorylated enzyme is much less active than the monophosphorylated and the unphosphorylated enzyme because of a decrease in its ability to bind calmodulin. The phosphatase also dephosphorylates the light-chain kinase; both phosphates are split in the absence of calmodulin, but only one in the presence of this protein. In the presence of calmodulin, only one site is phosphorylated in the kinase. Interestingly, this site is the site that cannot be phosphorylated in the calmodulin-kinase complex.

2. Phospholamban Kinase

Phospholamban is a small protein extractable by chloroform/methanol which is present in cardiac, but not in skeletal muscle sarcoplasmic reticulum membranes. Phospholamban is phosphorylated by Ca^{2+}-calmodulin-dependent kinases at a site which is different from that at which it is phosphorylated by cAMP-dependent protein kinases. Phosphorylation is associated with a several-fold increase in the rate of Ca^{2+} pumping.[54]

3. Protein Kinases of Sarcoplasmic Reticulum of Skeletal Muscle

A calmodulin-dependent kinase and probably also a calmodulin-dependent phosphatase have been detected in membranes from the sarcoplasmic reticulum of fast skeletal muscle. The kinases phosphorylate three proteins having M_r of 20,000, 35,000, and 57,000. The 20,000-dalton protein is an acidic proteolipid distinct from phospholamban. The physiological role of these kinases is unknown.[55]

D. Nervous Tissue Protein Kinases

The brain contains one type of cAMP-dependent and one type of cGMP-dependent protein kinase, but at least four types of Ca^{2+}-calmodulin dependent kinases have been found.[56] One of them, calmodulin kinase II, has been purified to homogeneity.[57] Brain protein kinases are involved in a multitude of metabolic phenomena including biosynthesis of neurotransmitters, neurotransmission, cell differentiation, ion channel conductances, etc.[58]

As will be discussed in more detail later, the entry of Ca^{2+} is the cause of the release of neurotransmitters at the synaptic terminal. The inflow of Ca^{2+} is followed by the phosphorylation of proteins of the presynaptic terminal through a reaction catalyzed by Ca^{2+}-calmodulin-dependent protein kinases.[58] Several proteins have been identified as endogenous substrates for these kinases. One is that designated initially by Greengard and co-workers[56] "protein I" and which now is called "synapsin I". This protein is present in all synaptic terminals as an extrinsic protein of the outer surface of the synaptic vesicles. Synapsin I belongs to a class of proteins with highly asymmetric structure and M_r about 86,000, it comprises about 0.4% of the total brain protein and is also a substrate for cAMP-dependent protein kinase. Experimental evidence has also been provided suggesting that Ca^{2+}-calmodulin-dependent protein kinases of synaptic terminals phosphorylate tubulin which upon phosphorylation changes its physical and chemical properties and aggregates, forming filamentous structures. These structures might interact with the presynaptic membranes and participate in the process of Ca^{2+}-mediated exocytosis, which is the basis of neurotransmitter release.[59,60]

The participation of calmodulin kinases in neurotransmission extends to other processes apart from the release of the transmitter. These include:

1.   Postsynaptic phenomena: for instance, calmodulin kinase II is a component of the major postsynaptic density protein[57] and Ca^{2+}-calmodulin-dependent kinases phosphorylate proteins in the postsynaptic membrane of the electric organ of the torpedo. It has been claimed that some are subunits of the acetylcholine receptor,[61] a finding that has not been confirmed by Huganir and Greengard[62] who postulate that in electric organ, the Ca^{2+}-calmodulin complex activates the phosphorylation of proteins that are independent of the subunits of the acetylcholine receptor.

2.   Biosynthesis of neurotransmitters: an example of this is the stimulation of brain tryptophan-5-monoxygenase by a calmodulin-dependent protein kinase. This enzyme catalyzes the initial and rate-limiting step in the biosynthesis of serotonin.[63,64]

The existence of protein kinases in nervous tissue implies the existence of protein phosphatases. One of the most conspicuous examples of this kind of enzyme is calcineurin whose properties have been analyzed when describing protein phosphatase 2B.

E. Membrane Protein Kinases in Other Tissues

Ca^{2+}-calmodulin-dependent protein kinase activity has been detected in the membrane fraction of all the tissues examined. These enzymes phosphorylate tissue-specific substrate proteins.[65] Although the phsyiological role of many of these enzymes is still unknown, it is likely that they mediate second-messenger actions of Ca^{2+}.

F. Plasma Membrane Ca²⁺-ATPases

As will be discussed in detail in Chapter 10, primary active transport systems that extrude Ca^{2+} from the cytosol towards the extracellular medium are activated by the Ca^{2+}-calmodulin complex. This is an interesting phenomenon from the point of view of regulation of intracellular signaling by Ca^{2+}, since it suggests that apart from initiating the response to the signal delivered by Ca^{2+}, Ca^{2+}-calmodulin complex also promotes a process that will eventually turn off the signal.

G. Other Enzymes
1. NAD Kinase

This enzyme catalyzes the phosphorylation of NAD to form NADP. The Ca^{2+}-dependence of this enzyme may be related to the role of Ca^{2+} in the synthesis of many cellular components, including steroids and nucleotides, and may also explain in part the role of Ca^{2+} during early development of the embryo. This enzyme has been detected in higher plants[66] which suggests that Ca^{2+} may also act as a second messenger in plant cells.

2. Platelet Phospholipase A2

This enzyme catalyzes hydrolysis of phospholipids that release arachidonic acid which in platelets is a precursor of thromboxane A2, a promoter of platelet aggregation.[67]

3. 15-Hydroxyprostaglandin Dehydrogenase

This enzyme initiates the first step of prostaglandin degradation resulting in the rapid loss of the biological activity of prostaglandins.[68]

V. Ca²⁺-DEPENDENT ENZYMES THAT DO NOT REQUIRE CALMODULIN

A. Protein Kinase C

This enzyme exists in an inactive state in the soluble fraction of all mammalian cells. In many tissues, the activity of this enzyme far exceeds that of cAMP-dependent protein kinase. The enzyme is a single polypeptide with no subunit structure and M_r 77,000. It requires phospholipids — in particular, phosphatidylserine — and it associates with the membrane to exhibit its catalytic activity. However, in the presence of phospholipids alone, it exhibits low affinity for Ca^{2+} (K_{Ca} 10 to 100 μM). When small amounts of diacylglycerols are also present, the apparent affinity for Ca^{2+} sharply increases and the enzyme becomes stimulated by concentrations of Ca^{2+} within the physiological range of cytosolic Ca^{2+}.[69-72] The diacylglycerol active in this role contains at least one unsaturated fatty acid such as arachinodate. Diacylglycerols are normally almost absent from cell membranes, but are transiently produced by the hydrolysis of phosphatidylinositol that follows the binding of many agonists to membrane receptors. In contrast with most of the other Ca^{2+}-dependent protein kinases, protein kinase C does not require calmodulin to be activated by Ca^{2+}. In spite of this, it is strongly inhibited by the so-called anticalmodulin agents as the phenotiazines.[70] The properties of protein kinase C suggest the existence of a general mechanism of signal transduction involving a transmembrane control of Ca^{2+}-dependent protein phosphorylation by protein kinase C which becomes activated as a consequent of a transient increase in the membrane concentration of diacylglycerol that follows the phosphatidylinositol response to stimuli.[69-71] Platelets provide an example of the possible physiological role of protein kinase C. In this system, release of serotonin after stimulation by trombin is associated with an increase in phosphatidylinositol turnover and with the phosphorylation by phosphokinase C of a protein of M_r 40,000. This reaction occurs concomitantly with the Ca^{2+}-calmodulin-dependent phosphorylation of another protein of M_r 20,000 and which has been identified with myosin light-chain.[73] The tumor-promoting agent tetradecanoylphorbol (TPA) substitutes for diacylglycerol as an activator of protein kinase C. TPA is much more potent than diacylglycerol, and it is not inactivated by the diacylglycerol kinase of activated cells. Diacylglycerol and TPA seem to activate the kinase in the same way and probably share the same binding sites in the enzyme.[70-73]

In connection with the mechanism of activation of phosphokinase C, it is interesting to consider the experiments by Rink et al.[74] These workers measured the time course of cytosolic Ca^{2+} concentration in platelets using quin2. They found that stimulation by exogenous diacylglycerol is not associated with any detectable change in cytosolic Ca^{2+} concentration although modest increases in cytosolic Ca^{2+} that are insufficient to stimulate secretion markedly enhance the response to diacylglycerol. This result suggests that, at least in some cases, diacylglycerol per se, rather than Ca^{2+}, has to be considered the second messenger for stimuli which elicit responses through phosphokinase C. In these cases, Ca^{2+} would exert a synergistic action.[69,71,72] Activation of the kinase by diacylglycerol may be either a direct effect of the lipid or the expression of the fact that in the presence of diacylglycerol, the enzyme can be stimulated by the physiological concentrations of Ca^{2+} in the cytosol of a cell at rest.

B. Calpain and Calpastatin

In 1981, Murachi[75] proposed the collective name of calpain to designate a family of endopeptidases that are located in the cytosol, of M_r 70,000 to 90,000, which are active only in the presence of Ca^{2+} and $-SH$ reducing agent. There seem to be at least two classes of calpains, one with high and the other with low affinity towards Ca^{2+}, which might result from the conversion by limited proteolysis of a single class of enzyme. However, even the high-affinity form of the enzyme has an apparent affinity for Ca^{2+} ($K_{Ca} = 10\ \mu M$) which is much lower than the affinity of other Ca^{2+}-binding proteins and about one order of magnitude higher than the physiological cytosolic Ca^{2+} concentration. Calpains are specifically inhibited by a family of endogenous protein inhibitors which are collectively designed as calpastatin. Calpastatins are large molecular weight proteins (M_r 250,000 to 300,000). They are specific for calpain, and do not inhibit trypsin, chymotrypsin, or papain. It has been proposed that the calpain-calpastatin system plays a regulatory role in parallel with the much better-known role of calmodulin in mediating intracellular effects of Ca^{2+}. Since the effects of calpain involve the hydrolysis of peptide bonds, this system, in contrast with the Ca^{2+}-calmodulin system, would mediate irreversible effects. However, before accepting this view, it is necessary to explain how physiological levels of Ca^{2+} would interact with a system of relatively low apparent affinity for the cation and which are the endogenous substrates for the proteolytic activity of the enzyme. (For an up-to-date review and reference on this system, see Reference 75).

VI. REGULATION OF CELL FUNCTIONS BY Ca²⁺

A. Endo- and Exocytosis

Present experimental evidence strongly suggests that Ca^{2+} has an important role in the processes that lead to the formation of endo- and exocytic vesicles. Although little is known about the detailed biochemistry of these processes, it seems that the effects of Ca^{2+} are mediated by the Ca^{2+}-calmodulin complex. Endo- and exocytic vesicles have, in their external regions, sites that bind calmodulin or Ca^{2+}-calmodulin. The effects of Ca^{2+} and calmodulin are, at least in some cases, mediated by the phosphorylation of membrane vesicles.

One of the best-studied processes of exocytosis is the release of neurotransmitters from the nerve terminal at synapsis. In these structures, the depolarization wave carried by the action potential opens voltage-dependent Ca^{2+} channels in the nerve terminal allowing the entry of extracellular Ca^{2+} with the consequent rise in cytosolic Ca^{2+}.[76,77] Ca^{2+} and calmodulin bind to the synaptic vesicles and initiate a series of biochemical and morphological events whose ultimate result is that vesicles fuse with the plasma membrane and release their contents into the space that separates the nerve endings and the postsynaptic region.[55,58,59] An essentially similar process probably occurs at the neuromuscular junction. An increase in Ca^{2+} concentration also triggers catecholamine secretion from chromaffin cells of the adrenal medulla. Calmodulin binds with high affinity to chromaffin granule proteins and this promotes the binding of other cytosolic proteins.[78]

There is also evidence for the participation of Ca^{2+} and calmodulin in exocytosis from nonneural cells. It is known that Ca^{2+} participates both in platelet aggregation and in the release through exocytosis of substances which cause vasoconstriction and facilitate rapid clotting of the blood[73,79] and that calmodulin interacts with the α-granules of platelets.[80] Another example of this apparently general phenomenon is provided by fertilized sea urchin eggs. In these cells during fertilization, secretory granules which lie directly beneath the plasma membrane fuse with the plasma membrane and expulse their contents into the extracellular space. This process is triggered by the elevated cytosolic Ca^{2+} concentration which can be observed during fertilization. Immunofluorescence studies of the localization of calmodulin show that it is present in the vicinity of the external surface of the granules at the region of their mutual attachment.[81] Release of hormones (see, for instance, References 82 and 83) also involve the participation of Ca^{2+} and calmodulin.

Both bulk phase and receptor-mediated endocytosis are blocked by the calmodulin antagonist TFP.[84] Coated vesicles from the brain have been shown to contain high-affinity calmodulin-binding sites.[85] The mechanism of the effects of calmodulin and Ca^{2+} during endocytosis is not known, but the presence of receptor sites for calmodulin in the surface of the vesicles may be related to their structural integrity and to their interactions with membrane and cytoskeleton components.

B. Cell Motility and Self-Assembling Cell Components

In contrast with cardiac and skeletal muscle, smooth muscle and nonmuscular cells contain actin and myosin, but they lack troponin C. In these tissues, the protein that mediates the effects of Ca^{2+} on contractile elements and on the cytoskeleton is calmodulin. We have already mentioned the key role of the Ca^{2+}-calmodulin-dependent myosin light-chain kinase (see Section IV.C. in this chapter). This enzymic reaction, however, is not the only mediator of the effects of Ca^{2+}. In smooth muscle, a calmodulin-binding protein called caldesmon has been isolated and identified.[86] This protein has the interesting property of binding to F actin when calmodulin has no Ca^{2+} bound and to calmodulin when it has its Ca^{2+} sites occupied. Binding of caldesmon to calmodulin prevents its association with F actin. The association of caldesmon with actin prevents the actin-myosin interaction. Hence caldesmon-calmodulin interactions acts as an on-off "switch" for the actin-myosin association. In smooth muscle, there seems to be another system which works through a similar mechanism. This system involves a protein of M_r 135,000 which at low (resting) Ca^{2+} levels binds to calmodulin, whereas at high Ca^{2+} concentration, it binds to F actin. This protein has myosin light-chain kinase activity and may be the enzyme itself. Therefore, the Ca^{2+}-calmodulin system not only triggers smooth muscle contraction by activation of myosin light-chain phosphorylation, but also seems to have a direct role in the association of myosin and actin filaments which is essential for contraction to take place (see References 87 and 88).

Spectrin is an oligomeric protein initially identified in the submembrane cytoskeleton of erythrocytes. Spectrin binds to a number of other proteins including F actin and intervenes in the control of the shape and in the linkage of contractile proteins to the plasma membrane of the erythrocyte. Recently, proteins which are structurally and functionally homologous to erythrocyte spectrin have been identified in brain and in other tissues. Both erythrocyte spectrin and the spectrin-like molecules of other cells are Ca^{2+}-dependent calmodulin-binding proteins.[88,89] Although the exact physiological role of spectrin-like molecules is not known, they are possibly implicated in the structural integrity of the plasma membrane, in the control of the mobility of membrane proteins in axoplasmic transport, and in the release of transmitters at the synapse. The ability of this class of proteins to associate with Ca^{2+}-calmodulin complexes suggests a role for this complex in the regulation of the functions of spectrin.[87-89]

Microtubules are components of the cytoskeleton of the cell which are associated with cellular processes involving motility, such as chromosome motion, neurite extension, and axonal transport. Depending on the conditions of the intracellular environment, microtubules assemble or disassemble by polymerization and depolymerization of their constitutents and, in particular, of the globular protein tubulin. Increases in the concentration of Ca^{2+} leads to the depolymerization of microtubules; calmodulin greatly decreases the concentration of Ca^{2+}, which induces this phenomenon. It has been shown that in the presence of Ca^{2+}, calmodulin forms a complex with tubulin which makes tubulin incapable of polymerization into microtubules. Another mechanism which probably is implied in the Ca^{2+}-calmodulin effect on the assembly of microtubules is that shown by Kakiuchi and co-workers (for references see References 87 and 88). These authors demonstrated that the so-called "tau" factor, a family of closely related low-molecular-weight proteins which are necessary for microtubule assembly, is a calmodulin-binding system. The interaction of the "tau" factor with calmodulin seems to follow an on-off switching mechanism very similar to that discussed previously for the case of caldesmon and F actin. In fact, at low Ca^{2+} concentrations when calmodulin is free of Ca^{2+}, the "tau" factor would interact with tubulin and promote the formation of microtubules. On the contrary, as Ca^{2+} concentration rises, the Ca^{2+}-calmodulin complex would bind to the "tau" factor, preventing its interaction with tubulin and thereby leading to the disassembly of microtubules. Microtubules that assemble in a crude brain extract contain a substantial subpopulation that is stable in the cold and resistant to drugs that inhibit the assembly of microtubules in vitro. It is interesting to note that in the mammalian mitotic apparatus, all the kinetocore-to-pole microtubules belong to the cold-stable type. Neither Ca^{2+} nor calmodulin alone have any effect on cold-stable microtubules, but disassembly is rapidly achieved by the simultaneous presence of both Ca^{2+} and calmodulin. Cold-stability seems to be conferred by a small number of polypeptides. The stabilizing activity of these is abolished by phosphorylation catalyzed by two kinds of protein kinases, one of which is dependent on Ca^{2+}-calmodulin. This kinase activity explains why at very low concentrations of calmodulin, destabilization of cold-stable microtubules becomes strictly dependent on the presence of ATP. These results, together with the particular distribution of cold-stable microtubules and with the preferential localization of calmodulin at the mitotic poles during cell division, suggest that Ca^{2+}-calmodulin-regulated diassembly of microtubules might promote chromosome movement towards the poles during anaphase and help in the establishment of a constant poleward microtubule treadmilling reaction during mitosis[90,91] (for a review and references on mitosis, see Reference 92).

C. Cell Division

The previously mentioned role of Ca^{2+} and calmodulin during mitosis focuses the attention on the involvement of Ca^{2+} in the initiation of cell division.[92] Mouse oocytes can be activated by the intracellular injection of Ca^{2+} and hamster oocytes by exposure to the Ca^{2+} ionophore A23187. It is known that in sea urchin eggs,[77] in the eggs of the medaka, the Japanese killifish,[94] and in mouse oocytes,[95] fertilization is associated with a sudden increase in the cytosolic Ca^{2+} level. It is likely that this phenomenon is involved during fertilization in most species. At least in sea urchin eggs, the increase in the concentration of Ca^{2+} triggers a series of responses among which we have already mentioned the promotion of exocytosis and the stimulation of the calmodulin-dependent enzyme NAD kinase. Ca^{2+} also plays an essential role during mitogenic stimuli in other cells. This has been studied to a certain detail in lymphocytes.[96] In these cells, mitogenic lectins and presumably more specific antigens bind to specific receptors at the cell surface and thus initiate a series of biochemical events that lead to mitosis. Binding of lectins requires extracellular Ca^{2+} but, more interestingly, it seems that the mitogenic response that follows this binding is triggered by an increase in cytosolic Ca^{2+}. The main evidences for this view are the following:

1. The mitogenic response does not take place in the absence of external Ca^{2+}.
2. Binding of lectins is associated with an increase in the rate of uptake of $^{45}Ca^{2+}$ from the external medium.
3. Studies with quin2 show that following a lectin stimulus, cytosolic Ca^{2+} doubles its concentration.[97]
4. The mitogenic effect of lectin can be mimicked by the Ca^{2+}-ionophore A23187.

Taken together, these results seem to indicate that one of the primary events that follows lectin binding to the surface of lymphocytes is the increase in the membrane permeability to Ca^{2+}. The mitogenic response to elevated intracellular Ca^{2+} is probably mediated by calmodulin, since this protein has been shown to be present in the cytosol of lymphocytes, and calmodulin antagonists also interfere with cell proliferation.

D. Membrane Channels
1. The Gárdos Effect
For a review on this subject, see Reference 98. In 1959, Gárdos showed that when red cells were incubated in the presence of metabolic inhibitors, they selectively lost a large amount of K^+.[99] The loss was not compensated by the entry of Na^+ and hence was accompanied by the shrinkage of the cells. A salient feature of this phenomenon is its absolute dependence on extracellular Ca^{2+}. In 1966, Schatzmann discovered the ATP-dependent active Ca^{2+} transport system in red cells (see Chapter 3). This allowed Whittam to suggest in 1968[100] that the selective loss of K^+ is the result of the build-up of the intracellular concentration of Ca^{2+} as a consequence of the slowing down of the mechanisms responsible for its active extrusion. Experimental results by Lew[101] and Romero and Whittam[102] demonstrated that Whittam's conjecture was correct. The study of the concentration dependence of the effects of internal Ca^{2+} in reconstituted ghosts loaded with Ca^{2+} buffers (Figure 5)[103] showed that K^+ loss started at near physiological (less than micromolar) concentrations of Ca^{2+} and hence it could have physiological importance. A Ca^{2+}-dependent increase in K^+ permeability has also been detected in heart muscle sarcolemma.[104] However, in this system, and in contrast with red cells, the apparent affinity for the Ca^{2+}-dependent increase in permeability to K^+ ($K_{Ca} = 10 \ \mu M$) is much higher than the physiological resting Ca^{2+} concentration of the tissue. However, the authors mention that Ca^{2+} concentrations as high as $10 \ \mu M$ can be reached in heart sarcoplasma during contractile activation.

Although it has been suggested that the Gárdos effect expresses an abnormal mode of behavior of the Na^+ pump,[105] the current view of this phenomenon is that intracellular Ca^{2+} opens a nonsaturable channel with high specificity towards K^+.[106] In the red cell, net K^+ movements through this channel are electrically balanced by the movement of small mobile anions such as chloride. When the effect is large, the permeability to anions may be a rate-limiting factor in the net movement of K^+.[106] The effect of Ca^{2+}

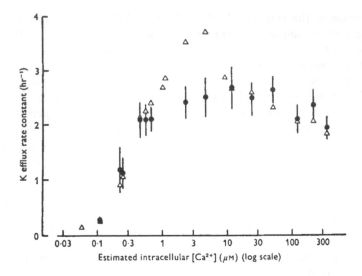

FIGURE 5.   The effect of intracellular Ca²⁺ on K⁺ efflux ("Gárdos effect") from human red blood cell ghosts containing 10 mM K⁺ and 90 mM choline (●) and 100 mM K⁺ (△).(From Simon, T. J. B., *J. Physiol.*, 256, 227, 1976. With permission.)

on K⁺ permeability can be blocked by quinine or quinidine.[97,107] Since K⁺ inhibits competitively the effect of these compounds, it is likely that quinine and quinidine combine directly at the K⁺ channel. Other drugs that inhibit the Gárdos effect are the phenothiazines and the diphenylbutyl piperidines which block this effect in red cell ghosts.[108] The effect of phenotiazine has also been observed in sarcolemmal vesicles of heart muscle.[104] Since these drugs inhibit numerous calmodulin-dependent phenomena, it is likely that, as with many other effects of intracellular Ca²⁺, it is not the cation per se, but the Ca²⁺-calmodulin complex that is the actual ligand that opens the K⁺ channel.

Although the Gárdos effect has been studied in greatest detail in red cells, it seems that intracellular Ca²⁺ controls the potassium permeability in many and perhaps in all cell membranes.[97,109-112] The physiological significance of the Gárdos effect in red cells, if any, is still obscure. In excitable cells, it may play a role in modulating the membrane potential after the passage of stimuli that increase the intracellular concentration of K⁺.

2. Intercellular Communication

Over the past 20 years, it has become clear that in cell systems whose components are not isolated in a liquid suspension, cells are communicated by relatively large (about 1.7 nm in diameter) channels across which hydrophilic molecules can flow directly from one cell interior to the other. These channels are located in the so-called gap junctions in which there is a close apposition of cell membranes (for references, see Reference 113). They are formed by several protein subunits whose components are contributed equally by the two cells that communicate. Cell-to-cell channels are found nearly everywhere in organized tissues throughout the phylogenetic scale. The intercellular channel seems to be essential for the coordinated behavior of a given tissue and provides a pathway for tissue homeostasis, for the transmission of electrical or chemical regulatory signals, and for the establishment of spatial and temporal patterns of physiological states and differentiation in multicellular systems. The state of opening or closure of intracellular communications is regulated by the concentration of intracellular Ca²⁺. As cytosolic Ca²⁺ rises, the channels pass from an open to a closed state. The transition between these two states takes place in the concentration range of 0.1 to 10 μM Ca²⁺. The transition is gradual, since fluxes of large molecules are slowed at lower Ca²⁺ concentrations than fluxes of small molecules. It is not yet known whether this gradual effect represents a gradual closure of a single population of channels or an all-or-none closing of an heterogenous population of channels. The role of calmodulin in this process is still not clear. The main physiological reason for the Ca²⁺-dependent closing of the channels is that it provides a mechanism that allows an interconnected group of cells to seal themselves from an unhealthy member since, as has been pointed out by Loewenstein,[113] the resting level of cytosolic Ca²⁺ is a good thermometer of the health of a cell.

REFERENCES

1. Kretsinger, R. H., Mechanism of selective signaling by calcium, *Neurosci. Res. Program Bull.*, 19, 217, 1981.
2. Urry, D. W., Basic aspects of calcium chemistry and membrane interaction: on the messenger role of calcium *Ann. N.Y. Acad. Sci.*, 307, 3, 1978.
3. Urry, D. W., Trapane, T. L., Walker, J. T., and Prasad, K. U., On the relative lipid membrane permeability of Na⁺ and Ca²⁺. A physical basis for the messenger role of Ca²⁺, *J. Biol. Chem.*, 257, 6649, 1982.
4. Kretsinger, R. H., Mechanism of selective signaling by calcium, *Neursoci. Res. Program Bull.*, 19, 297, 1981.
5. Somlyo, A. P., Cellular site of calcium regulation, *Nature*, 309, 516, 1984.
6. Reuter, H., Calcium channel modulation by neurotransmitters, enzymes and drugs, *Nature*, 301, 569, 1983.
7. Gould, R. J., Murphy, K. M. M., and Snyder, S. H., ³H-Nitrendipine-labelled calcium channels discriminate inorganic calcium agonists and antagonists. *Proc. Natl. Acad. Sci. U.S.A.*, 79, 3656, 1982.
8. Curtis, B. M. and Catterall, W. A., Solubilization of the calcium antagonist receptor from rat brain, *Proc. Natl. Acad. Sci. U.S.A.*, 258, 7280, 1983.
9. Nomrman, R. I., Borsotto, M., Fosset, M., Lazdunzki, M., and Ellory, J. C., Determination of the molecular size of the nitrendipine-sensitive Ca²⁺ channel by radiation inactivation, *Biochim. Biophys. Res. Commun.*, 111, 878, 1983.
10. Schramm, M., Thomas, G., Towart, R., and Franckowiak, Novel dihydropyridines with positive inotropic action through activation of Ca²⁺ channels, *Nature*, 303, 535, 1983.
11. Michell, R. H., Inositol phospholipids and cell surface receptor function, *Biochim. Biophys. Acta*, 415, 81, 1975.
12. Michell, R. H., Inositol phospholipids in membrane function, *TIBS*, 128, 1979.
13. Hawthorne, J. N., Is phosphatidylinositol now out of the calcium gate?, *Nature*, 295, 281, 1982.
14. Michell, R. H., Is phosphatidylinositol really out of the calcium gate?, *Nature*, 296, 492, 1982.
15. Hesketh, R., Inositol triphosphate: link or liability, *Nature*, 306, 16, 1983.
16. Streb, H., Irvine, R. F., Berrige, M. J., and Schultz, I., Release of Ca²⁺ from nonmitochondrial intracellular stores in pancreatic acinar cells, by inositol-1,4,5-triphosphate, *Nature*, 306, 67, 1983.
17. Suresh, K. J., Thomas, A. P., Willams, R. J., Irvine, R. F., and Willamson, J. R., Myo-inositol 1,4,5-triphosphate. A second messenger for the hormonal mobilization of intracellular Ca²⁺ in the liver, *J. Biol. Chem.*, 259, 3077, 1984.
18. Willamson, J. R., Cooper, R. H., and Hoek, J. B., Role of calcium in the hormonal regulation of liver metabolism, *Biochim. Biophys. Acta*, 639, 243, 1981.
19. Kretsinger, R. H., Calcium-binding proteins, *Ann. Rev. Biochem.*, 45, 916, 1976.
20. Kretsinger, R. H., Calcium-modulated proteins, *Neurosci. Res. Program Bull.*, 19, 226, 1981.
21. Kretsinger, R. H. and Barry, C. D., The predicted structure of the calcium-binding component of troponin, *Biochim. Biophys. Acta*, 405, 40, 1975.
22. Bredderman, P. J. and Wasserman, R. H., Chemical composition, affinity for calcium and some related properties of vitamin D dependent calcium-binding protein, *Biochemistry*, 13, 1687, 1974.
23. Jande, S. J., Maler, L., and Lawson, D. E., Immunohistochemical mapping of vitamin D-dependent calcium-binding protein in brain, *Nature*, 294, 765, 1981.

24. Bryant, D. T. W. and Andrews, P., A simple procedure for purifying mammalian duodenal Ca²⁺-binding proteins on a 100-mg scale and an investigation of the stoichiometry of their high affinity binding of Ca²⁺ ions, *Biochem. J.,* 211, 709, 1983.

25. Cheung, W. Y., Cyclic 3',5'-nucleotide phosphodiesterase. Demonstration of an activator, *Biochim. Biophys. Res. Commun.,* 38, 533, 1970.

26. Kakiuchi, S. and Yamazaki, R., Calcium dependent phosphodiesterase activity and its activating factor (PAF) from brain. III. Studies on cyclic 3'-5'-monophosphate phosphodiesterase, *Biochim. Biophys. Res. Commun.,* 41, 1104, 1970.

27. Teo, T. S. and Wang, J. H., Mechanism of the activation of a cyclic adenosine 3':5'-monophosphate phosphodiesterase from bovine heart by calcium ions. Identification of the protein activator as a Ca²⁺-binding protein, *J. Biol. Chem.,* 248, 5950, 1973.

28. Cheung, W. Y., Lynch, T. J., and Wallace, R. W., An endogenous Ca²⁺-dependent activator protein of brain adenylate kinase and brain phosphodiesterase, *Adv. Cyclic Nucleotide Res.,* 9, 233, 1978.

29. Cheung, W. Y., Calmodulin plays a pivotal role in cellular regulation, *Science,* 207, 19, 1979.

30. Klee, C. B., Crouch, T. H., and Richman, P. G., Calmodulin, *Ann. Rev. Biochem.,* 49, 489, 1980.

31. Means, A. R. and Dedman, J. R., Calmodulin — an intracellular calcium receptor, *Nature,* 285, 73, 1980.

32. Scharff, O., Calmodulin and its role in cellular activation, *Cell Calcium,* 2, 1, 1981.

33. Oldham, S. B., Calmodulin: its role in calcium-mediated cellular regulation, *Mineral Electrolyte Metab.,* 8, 1, 1982.

34. Cheung, W. Y., Calmodulin: an overview, *Fed. Proc. Fed. Am. Soc. Exp. Biol.,* 41, 2253, 1982.

35. Watterson, D. M., Sharief, F., and Vanaman, T. C., The complete amino acid sequence of Ca²⁺-dependent activator protein (calmodulin) of bovine brain, *J. Biol. Chem.,* 255, 462, 1980.

36. Ikura, M., Hiraoki, T., Hikichi, K., Mikuni, T., Yazawa, M., and Yagi, K., Nuclear magnetic resonance studies on calmodulin: spectral assignments in the calcium-free state, *Biochemistry,* 22, 2569, 1983.

37. Ikura, M., Hiraoki, T., Hikichi, K., Mikuni, T., Yazawa, M., and Yagi, K., Nuclear magnetic resonance studies on calmodulin: calcium-induced conformational changes, *Biochemistry,* 22, 2573, 1983.

38. Wang, C. H., Mutus, B., Sharma, R. K., Hing-Yat, P. L., and Wang, J. G., On the mechanism of interaction between calmodulin and calmodulin-dependent proteins, *Can. J. Biochem. Cell Biol.,* 61, 911, 1983.

39. LaPorte, D. C., Weirman, B. W., and Storm, D. R., Calcium-induced exposure of a hydrophobic surface on calmodulin, *Biochemistry,* 19, 3814, 1980.

40. Jamieson, G. A., Jr. and Vanaman, T. C., Calcium-dependent affinity chromatography of calmodulin on an immobilized phenotiazine, *Biochim. Biophys. Res. Commun.,* 90, 1048, 1979.

41. Charboneau, H. and Cormier, M. J., Purification of plant calmodulin by fluphenazine-Sepharose affinity chromatography, *Biochem. Biophys. Res. Commun.,* 90, 1039, 1979.

42. Cohen, P., The role of protein phosphorylation in neural and hormonal control of cellular activity, *Nature,* 296, 613, 1982.

43. Cohen, P., Protein phosphorylation and the control of glycogen metabolism in skeletal muscle, *Phil. Trans. R. Soc. London,* B302, 13, 1983.

44. Payne, E. M. and Soderling, T. R., Calmodulin-dependent glycogen synthethase kinase, *J. Biol. Chem.,* 255, 8054, 1980.

45. Woodgett, J. R., Davison, M. T., and Cohen, P., The calmodulin dependent glycogen synthethase kinase from rabbit skeletal muscle: purification, subunit structure and substrate specificity, *Eur. J. Biochem.,* 1983, in press.

46. McGuiness, T. L., Lai, Y., Greengard, P., Woodgett, J. R., and Cohen, P., A multifunctional calmodulin-dependent protein kinase: similarities between skeletal muscle glycogen synthethase kinase and a brain synapsin I kinase, *FEBS Lett.,* 1983, in press.

47. Ingebristen, T. S. and Cohen, P., The protein phosphatases involved in cellular regulation. I. Classification and substrate specificities, *Eur. J. Biochem.,* 132, 255, 1983.

48. Stewart, A., Ingebristen, T. S., and Cohen, P., The protein phosphatases involved in cellular regulation. V. Purification and properties of a Ca²⁺/calmodulin-dependent protein phosphatase (2B) from rabbit skeletal muscle, *Eur. J. Biochem.,* 132, 289, 1983.

49. Tallant, E. A. and Cheung, W. Y., Calmodulin-dependent protein phosphatase: a developmental study, *Biochemistry,* 22, 3630, 1983.

50. Manalan, A. S. and Klee, C. B., Activation of calcineurin by limited proteolysis, *Proc. Natl. Acad. Sci. U.S.A.,* 80, 4291, 1983.

51. Dabrowska, R., Sherry, J. M. F., Aramatoria, D. K., and Hartshorne, D. J., Modulator protein as a component of myosin light chain kinase from chicken gizzard, *Biochemistry,* 17, 253, 1978.

52. Yagi, K., Yazawa, M., Kakiuchi, S., Oshima, M., and Uenishi, K., Identification of an activator protein for myosin light chain kinase as Ca²⁺-dependent modulator protein, *J. Biol. Chem.,* 253, 1338, 1978.

53. Pato, M. D. and Adelstein, R. S., Purification and characterization of a multisubunit phosphatase from turkey gizzard smooth muscle. The effect of calmodulin binding to myosin light chain kinase on dephosphorylation, *J. Biol. Chem.*, 258, 7047, 1983.

54. LePeuch, C. J., LePeuch, D. A. M., and Demaille, J. G., Ca^{2+}-calmodulin-dependent phospholamban kinase from cardiac sarcoplasmic reticulum is distinct from phospholamban kinase and forms a regulatory complex with phospholamban and the Ca^{2+}-ATPase, *Ann. N.Y. Acad. Sci.*, 402, 549, 1982.

55. Chiese, M. and Carafoli, E., The role of calmodulin in skeletal muscle sarcoplasmatic reticulum, *Biochemistry*, 22, 985, 1983.

56. Nestler, E. J. and Greengard, P., Protein phosphorylation in the brain, *Nature*, 305, 583, 1983.

57. Kelly, P. T., McGuinness, T. L., and Greengard, P., Evidence that the major postsynaptic density protein is a component of Ca^{2+}-calmodulin-dependent protein kinase, *Proc. Natl. Acad. Sci. U.S.A.*, 81, 945, 1984.

58. Shulman, H. and Greengard, P., Stimulation of brain membrane protein phosphorylation by calcium and an endogenous heat-stable protein factor, *Nature*, 271, 478, 1978.

59. DeLorenzo, R. J., The calmodulin hypothesis of neurotransmission, *Cell Calcium*, 2, 365, 1981.

60. DeLorenzo, R. J., Calmodulin in neurotransmitter release and synaptic function, *Fed. Proc. Fed. Am. Soc. Exp. Biol.*, 41, 2265, 1982.

61. Smilowitz, H., Hadjian, R. A., Dwyer, J., and Feinstein, M. B., Regulation of acetylcholine receptor phosphorylation by calcium and calmodulin, *Proc. Natl. Acad. Sci. U.S.A.*, 78, 4709, 1981.

62. Huganir, R. L. and Greengard, P., cAMP-dependent protein kinase phosphorylates the nicotinic acetylcholine receptor, *Proc. Natl. Acad. Sci. U.S.A.*, 80, 1130, 1983.

63. Kuhn, D. M., O'Callahan, J. P., Juskevich, J., and Lovenberg, W., Activation of brain triptophan hydroxylase by ATP-Mg^{2+}: dependence on calmodulin, *Proc. Natl. Acad. Sci. U.S.A.*, 77, 4688, 1980.

64. Kuhn, D. M. and Lovenberg, W., Role of calmodulin in the activation of tryptophan hydroxylase, *Fed. Proc. Fed. Am. Soc. Exp. Biol.*, 41, 2258, 1982.

65. Shulman, H. and Greengard, P., Ca^{2+}-dependent protein phosphorylation system in membranes from various tissues and its activation by "calcium-dependent" regulator, *Proc. Natl. Acad. Sci. U.S.A.*, 75, 5432, 1978.

66. Anderson, J. M., Charbonneau, H., Jones, H. P., McCann, R. O., and Cormier, M. J., Characterization of plant nicotinamide adenine dinucleotide kinase activator protein and its identification as calmodulin, *Biochemistry*, 19, 3113, 1980.

67. Wong, P. Y. K. and Cheung, W. Y., Calmodulin stimulates human platelet phospholipase A2, *Biochim. Biophys. Commun.*, 90, 473, 1979.

68. Wong, P. Y. K., Lei, W. H., Chao, P. H. W., and Cheung, W. Y., The role of calmodulin on prostaglandin metabolism, *Ann. N.Y. Acad. Sci.*, 356, 179, 1980.

69. Michell, B., Ca^{2+} and protein kinase C: two synergistic cellular signals, *TIBS*, 8, 263, 1983.

70. Niskhizuka, Y., The role of protein kinase C in cell surface signal transduction and tumor promotion, *Nature*, 308, 693, 1984.

71. Nisishuka, Y., Protein kinases in signal transduction, *TIBS*, 9, 163, 1984.

72. Kishimoto, A., Takai, Y., Terutushi, M., Kikkawa, U., and Nishizuka, N., Activation of calcium and phospholipid-dependent protein kinase by diacylglycerol. Its possible relation to phosphatidyl inositol turnover, *Proc. Natl. Acad. Sci. U.S.A.*, 256, 2277, 1980.

73. Kaibuchi, K., Takai, Y., Sawamura, M., Hoshijima, M., Fujikura, T., and Nishizuka, Y., Synergystic functions of protein phosphorylation and calcium mobilization in platelet activation, *J. Biol. Chem.*, 258, 6701, 1983.

74. Rink, T. J., Sanchez, A., and Hallam, T. J., Diacylglycerol and phorbol ester stimulate secretion without raising cytoplasmic free calcium in human platelets, *Nature*, 305, 317, 1983.

75. Murachi, T., Intracellular Ca^{2+} protease and its inhibitor protein: calpain and calpastain, *Calcium and Cell Function*, 4, 377, 1983.

76. Llinas, R., Sugimori, M., and Simon, S. M., Transmission by presynaptic spike-like depolarization in the squid giant synapse, *Proc. Natl. Acad. Sci. U.S.A.*, 79, 2415, 1982.

77. Llinas, R., Steimberg, I. Z., and Walton, K., Relationship between presynaptic calcium current and postsynaptic potential in squid giant synapse, *Biophys. J.*, 33, 232, 1981.

78. Geisow, M. J. and Burgoyne, R. D., Recruitment of cytosolic proteins to a secretory granule membrane depends on Ca^{2+}-calmodulin, *Nature*, 301, 432, 1983.

79. Rink, T. J., Smith, S. W., and Tsien, R. Y., Cytoplasmic free Ca^{2+} in human platelets: Ca^{2+} thresholds and Ca-independent activation for shape change and secretion, *FEBS Lett.*, 148, 21, 1982.

80. Grinstein, S. and Furuya, W., Calmodulin binding to platelet plasma membranes, *FEBS Lett.*, 140, 49, 1982.

81. Steinhardt, R. A. and Alderton, J. M., Calmodulin confers calcium sensitivity on secretory exocytosis, *Nature*, 295, 154, 1983.

82. Schubart, U. K., Erlichman, J., and Fleischer, N., Insulin release and protein phosphorylation: possible role of calmodulin, *Fed. Proc. Fed. Am. Soc. Exp. Biol.*, 41, 2278, 1982.

83. Colca, J. R., Kotaga, C. L., Lacy, P. E., Landt, M., and McDaniel, M. L., Alloxan inhibition of Ca^{2+}-and-calmodulin-dependent protein kinase activity in pancreatic islets, *J. Biol. Chem.*, 258, 7260, 1983.

84. Salisbury, J. L., Condeelis, J. S., and Satir, P., The role of coated vesicles, microfilaments and calmodulin in receptor mediated endocytosis by cultured B lymphoblastoid cells, *J. Cell. Biol.*, 87, 132, 1980.

85. Linden, C. D., Dedman, J. R., Chafouleas, J. G., Menas, A. R., and Roth, T. F., Interactions of calmodulin with coated vesicles from brain, *Proc. Natl. Acad. Sci. U.S.A.*, 78, 308, 1981.

86. Sobue, K., Muramoto, Y., Fujita, M., and Kakiuchi, S., Purification of a calmodulin-binding protein from chicken gizzard that interacts with F-actin, *Proc. Natl. Acad. Sci. U.S.A.*, 78, 5652, 1981.

87. Kakiuchi, S. and Sobue, K., Control of cytoskeleton by calmodulin and calmodulin-binding proteins, *TIBS*, 8, 59, 1983.

88. Sobue, K., Kanda, K., Adachi, J., and Kakiuchi, S., Calmodulin-binding proteins that interact with actin filaments in a Ca^{2+}-dependent flip-flop manner: survey in brain and secretory tissues, *Proc. Natl. Acad. Sci. U.S.A.*, 80, 6668, 1983.

89. Sobue, K., Muramoto, Y., Fujita, M., and Kakiuchi, S., Calmodulin-binding protein in erythrocyte cytoskeleton, *Biochim. Biophys. Res. Commun.*, 100, 1063, 1981.

90. Job, D., Fischer, E. H., and Margolis, R. L., Rapid disassembly of cold-stable microtubules by calmodulin, *Proc. Natl. Acad. Sci. U.S.A.*, 78, 4679, 1981.

91. Job, D., Rauch, C. T., Fischer, E. H., and Margolis, R. L., Regulation of microtubule cold stability by calmodulin-dependent and independent phosphorylation, *Proc. Natl. Acad. Sci. U.S.A.*, 80, 3894, 1983.

92. McIntosh, J. R., Mechanisms of mitosis, *TIBS*, 9, 195, 1984.

93. Steinhardt, R. A., Epel, D., Carrol, E. J., and Yanagimachi, R., Is calcium ionophore a universal activator for unfertilized eggs?, *Nature*, 252, 41, 1974.

94. Ridgway, E. B., Gilkey, J. C., and Jaffe, L. F., Free calcium increases explosively in activating medaka eggs, *Proc. Natl. Acad. Sci. U.S.A.*, 74, 632, 1977.

95. Cuthberston, K. S. R., Whittingham, D. G., and Cobbold, P. H., Free Ca^{2+} increases in exponential phases during mouse oocyte activation, *Nature*, 294, 754, 1981.

96. Lichtman, A. H., Segel, G. B., and Lichtman, M. A., The role of calcium in lymphocyte proliferation (an interpretative review), *Blood*, 61, 413, 1983.

97. Tsien, R. Y., Pozzan, T., and Rink, T. J., T-cell mitogens cause early changes in cytoplasmic free Ca^{2+} and membrane potential in lymphocytes, *Nature*, 295, 68, 1982.

98. Lew, V. L. and Ferreira, H. G., Calcium transport and the properties of a calcium-activated potassium channel in red cell membranes, *Curr. Top. Membr. Transp.*, 10, 217, 1978.

99. Gardos, G., The role of calcium on the potassium permeability of human erythrocytes, *Acta Physiol. Acad. Sci. Hung.*, 15, 121, 1959.

100. Whittam, R., Control of membrane permeability to potassium in red blood cells, *Nature*, 219, 610, 1968.

101. Lew, V. L., Effect of intracellular calcium on the potassium permeability of human red cells, *J. Physiol. (London)*, 206, 35P, 1970.

102. Romero, P. J. and Whittam, R., The control by internal calcium of the membrane permeability to sodium and potassium, *J. Physiol. (London)*, 214, 481, 1971.

103. Simons, T. B. J., Calcium-dependent potassium exchange in human red cell ghosts, *J. Physiol. (London)*, 256, 227, 1976.

104. Caroni, P. and Carafoli, E., Modulation by calcium of the potassium permeability of dog heart sarcolemmal vesicles, *J. Biol. Chem.*, 79, 5763, 1982.

105. Blum, R. M. and Hoffman, J. F., Carrier mediation of Ca-induced K transport and its inhibition in red blood cells, *Fed. Proc. Fed. Am. Soc. Exp. Biol.*, 29, 633a, 1970.

106. Schubert, A. and Sarkadi, B., Kinetic studies on the calcium-dependent potassium transport in human red blood cells, *Acta Biochim. Biophys. Acad. Sci. Hung.*, 12, 207, 1077.

107. Reichstein, E. and Rothstein, A., Effects of quinine on Ca^{++}-induced K$^+$ efflux from human red blood cells, *J. Membr. Biol.*, 59, 57, 1981.

108. Lackington, I. and Orrego, F., Inhibition of calcium-activated potassium conductance of human erythrocytes by calmodulin inhibitory drugs, *FEBS Lett.*, 133, 103, 1981.

109. Burgess, G. M., Claret, M., and Jenkinson, D. H., Effects of quinine and apamin on the calcium-dependent potassium permeability of mammalian hepatocytes and red cells, *J. Physiol. (London)*, 317, 67, 1981.

110. Maruyama, Y., Petersen, O. H., Flanagan, P., and Pearson, G. T., Quantification of Ca^{2+}-activated K$^+$ channels under hormonal control in pig pancreas acinar cells, *Nature*, 305, 228, 1983.

111. Pallota, B. S., Magleby, K. L., and Barret, J. N., Single channel recordings of Ca^{2+}-activated potassium currents in rat muscle cell culture, *Nature,* 293, 471, 1981.

112. Marty, A., Ca-dependent K channels with large unitary conductances in chromaffin cell membranes, *Nature,* 291, 497, 1981.

113. Loewenstein, W. R., Junctional intercellular communication: the cell-to-cell membrane channel, *Physiol. Rev.,* 61, 829, 1981.

Chapter 4

FROM THE DISCOVERY OF THE Ca²⁺ PUMP IN PLASMA MEMBRANES TO THE DEMONSTRATION OF ITS UBIQUITOUSNESS

A. F. Rega

I. A HISTORICAL REVIEW

A. The First Report

In 1966, Hans J. Schatzmann reported the discovery of the Ca^{2+} pump of plasma membranes in a short paper entitled "ATP-dependent Ca^{++}-Extrusion from Human Red Cells," published in *Experientia*, Volume 22, page 364. At that time, after the work of Dunham and Glynn,[1] it was known that 100 μM Ca^{2+} inhibited the Na^+ pump of human red cells. The concentration of Ca^{2+} in the blood plasma is 3×10^{-3} *M.* In view of this, Schatzmann reasoned that "for obvious reasons the Ca^{++}-sensitive site must be located on the internal surface of the membrane and therefore the intracellular Ca^{++} concentration must be considerably lower than the Ca^{++} concentration in the plasma or the Na-K-pump would be incapacitated". Since Dunham and Glynn had also reported that the red cell membrane possesses a Ca^{2+}-dependent ATPase,[1] Schatzmann hypothesized that this ATPase could be connected to active extrusion of Ca^{2+} from the red cell, as the Ca^{2+}-dependent ATPase from sarcoplasmic reticulum is related to the accumulation of Ca^{2+} from the cytosol into this structure.

Schatzmann approached the question by measuring the changes in calcium concentration that occurred when resealed ghosts from human red cells were incubated at 37°C. The ghosts contained $CaCl_2$ and KCl as the major osmotic constituent and were suspended in isotonic medium containing $CaCl_2$ and NaCl buffered with Tris-HCl. Measuring by complexometric titration the increase in calcium concentration in the external fluid in which resealed ghosts containing either 0 or 4 mM ATP were incubated, Schatzmann found the results of Figure 1 and informed that "Ca emerged from ATP-loaded ghosts at a high rate and that the release proceeded to a higher external concentration than with control ghosts treated in the same way except for the addition of Mg-ATP." The next experimental result shown by Schatzmann in this first report was that of an experiment designed to compare the changes in the concentration of calcium in the medium and in the ghosts, after resealed ghosts containing 0 and 2 mM ATP were incubated at 37°C for 60 min. The results of such an experiment are in Figure 2. The author's comments were " . . . that this ATP-dependent loss of Ca (referring to that shown in Figure 2) resulted in a reversal of the original gradient, the final internal Ca concentration being considerably lower than the concentration for equal distribution of Ca between inside and outside . . . " "The determination of the chloride distribution under identical conditions showed that the membrane potential in the ghosts was similar to that in intact cells, i.e., the interior was negative by 5 to 8 mV. Therefore, passive Ca^{++} distribution should lead to an even higher internal Ca^{2+} concentration than that calculated on the assumption of equal distribution."

"From these facts it is clear that, in the presence of ATP inside the ghosts, Ca^{++}-movement against the electrochemical gradient took place during the incubation which was not observed in the absence of ATP."

Following this statement, Schatzmann argued against the possibility that the movement of Ca^{2+} observed in the presence of ATP could be due to either the accumulation of orthophosphate in the ghosts, the Na^+ and K^+ gradients, or the functioning of the

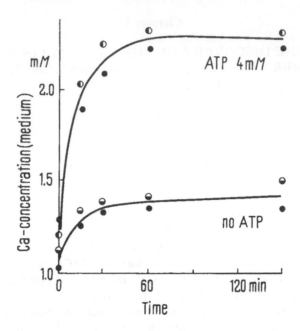

FIGURE 1.   Appearance of calcium in the external fluid at 37°C from ghosts of human red blood cells resealed with and without 4 mM ATP. (From Schatzmann, H. J., *Experientia*, 22, 364, 1966. With permission.)

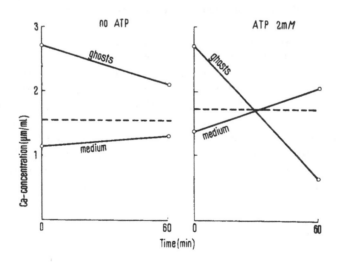

FIGURE 2.   The Ca concentration in ghosts with and without 2 mM ATP and in the medium after incubation at 37°C. The dotted lines represent the concentration of calcium calculated for equal distribution between ghosts and medium. (From Schatzmann, H. J., *Experientia*, 22, 364, 1966. With permission.)

Na⁺ pump, and after that, concluded, "These observations suggest that human red cells are able to maintain low intracellular Ca⁺⁺-concentration by aid of an active transport mechanism for this cation which derives its energy from ATP-splitting and operates independently from the Na-K-pump mechanism."

The hypothesis made by Schatzmann on the existence of a mechanism for the active transport of Ca²⁺ shows acute insight and its elegant demonstration represents a masterly piece of experimental work. Later on, Schatzmann and Vicenzi[2] confirmed and extended the experimental evidence on the existence of a Ca^{2+} pump in the erythrocyte membrane, demonstrating also that when transported, Ca^{2+} activated the hydrolysis of ATP at the internal surface of the membrane. These two reports represent the starting point of the knowledge that has accumulated on the Ca^{2+} pump from plasma membrane.

B. The Finding of the Ca^{2+} Pump in Plasma Membranes Other Than the Erythrocyte

It was not until recently that the existence of an ATP-driven active transport of Ca^{2+} in the plasma membrane of most cells was demonstrated. At the time Schatzmann discovered the Ca^{2+} pump in red cells, Reuter and Seitz proposed that in heart muscle cells, the energy for extruding Ca^{2+} across the plasma membrane is provided by the movement of Na^+ into the cell from the more concentrated solution outside, through a Na^+-Ca^{2+} countertransport system; Baker, Blaunstein, Hodgkin, and Steinhardt collected evidences for a similar mechanism in squid axon (for references, see Reference 3). The existence of a mechanism capable of extruding Ca^{2+} from excitable cells against electrochemical gradients utilizing the energy provided by the Na^+ concentration gradient generated by the Na^+ pump made it unnecessary for several years to postulate an ATP-driven Ca^{2+} pump in these cells.

In 1976, J. D. Robinson[4] reported that rat brain microsomes showed a Ca^{2+}-dependent ATPase activity and an ATP-dependent Ca^{2+} accumulation. After centrifugation in a sucrose gradient, the Ca^{2+}-ATPase activity was distributed together with the (Na^+-K^+)-ATPase activity and not with mitochondrial markers. After this finding, Robinson suggested that this ATPase may represent a Ca^{2+}-transport system across the neural plasma membrane. At the same time, C. J. Duncan[5] reported in synaptic membranes from rat brain an ATPase with high affinity for Ca^{2+} and an activation energy for the reaction of ATP hydrolysis in the range of those other transport ATPases. He concluded that the ATPase had the typical properties associated with a high-affinity Ca^{2+} transport system. Unfortunately, neither of the two authors showed in their reports experimental evidence to prove the hypothetical participation of the activities they described in the transport of Ca^{2+} across the plasma membrane. Furthermore, in 1978, Robinson[6] reported that his preparation from rat brain was also phosphorylated by ATP, and SDS gel electrophoresis of the phosphorylated preparation shows that Ca^{2+} stimulates incorporation into a single major fraction with a M_r of 100,000 instead of 150,000 as with the Ca^{2+}-ATPase from plasma membrane. However, this could well have been a product of proteolysis or maltreatment of the plasma membrane Ca^{2+}-ATPase, as has been suggested by Hakin et al.[7]

In 1978, after he had observed that most of the experiments describing the Na^+-Ca^+ exchange in squid axon had been carried out at Ca^{2+} concentrations exceedingly higher than the physiological one,[8] DiPolo[9] performed experiments on internally dialyzed squid axons in which Ca^{2+}_i was set at 0.45 μM with the use of EGTA. Under these conditions, addition of internal ATP increased Ca^{2+} efflux without any observable changes in Ca^{2+} influx (Figure 3), the result being a net Ca^{2+} efflux. Replacement of all the external Na^+ by Tris reduced this efflux by 35%, whereas removal of ATP reduced the efflux to the leakage values. ATP, Mg^{2+}, and Ca^{2+} exert their effects from the cytosol, because they were without effect when applied externally. These data constitute the first direct demonstration of an ATP-dependent mechanism that transports Ca^{2+} with high affinity across a plasma membrane from a cell different than the red cell. Since the squid axon does not provide one with enough material for the assays, DiPolo's report did not give experimental evidence that ATP undergoes hydrolysis when Ca^{2+} is transported.

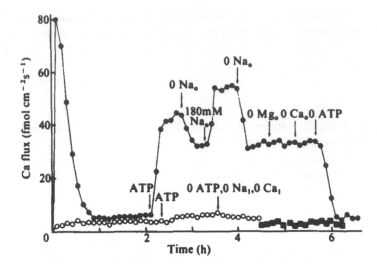

FIGURE 3.   Ca²⁺ influx (○), Ca efflux (●), and CaEGTA leakage (■) from internally dialyzed squid axons. (From DiPolo, R., *Nature*, 274, 390, 1978. With permission.)

Although without giving conclusive evidence on the existence of a Ca²⁺ pump in the plasma membrane of liver cells, in the same year that DiPolo reported his finding in axon, Cittadini and van Rossum[10] reported studies on the properties of the calcium-extruding mechanism of liver cells using slices of rat liver and concluded that the extrusion of Ca²⁺ from those cells was a metabolically dependent transport process that occurs independently of the exchange of Na⁺ between tissue and medium. One year later, Gmaj, Murer, and Kine,[11] using vesicles of basolateral plasma membrane from rat kidney cortex, demonstrated an ATP-dependent transport of Ca²⁺ into the vesicles distinguishable from the Ca²⁺-uptake into mitochondria or endoplasmic reticulum with K_{Ca} 0.5 μM.

The existence of ATP-dependent Ca²⁺ transport for the extrusion of Ca²⁺ in excitable cells was also demonstrated in dog heart sarcolemma by Caroni and Carafoli[12] in 1980 (Figure 4). These authors used a preparation highly enriched in sarcolemma vesicles and demonstrated that the vesicles hydrolyze ATP and accumulate Ca²⁺ in the absence of a Na⁺ gradient. The K_{Ca} for Ca²⁺ transport was 0.2 to 0.6 μM. Caroni and Carafoli[12] discarded the involvement of mitochondria in the accumulation of Ca²⁺, because neither oligomycin nor the uncoupler FCCP had any effect. Furthermore, the Ca²⁺ inside the vesicles was rapidly discharged by Na⁺ from the external media (Figure 4), ruling out the possibility that the ATP-dependent uptake of Ca²⁺ could be due to sarcoplasmic reticulum vesicles, since the sarcoplasmic reticulum membrane lacks a Na⁺-Ca²⁺ exchange mechanism. The authors concluded that the sarcolemma of heart cells possesses, in parallel with the Na⁺-Ca²⁺ exchange an ATP-dependent system which transports Ca²⁺. By that time, Morcos and Drummond[13] demonstrated a Ca²⁺-ATPase activity with a high affinity for Ca²⁺ in an enriched sarcolemma fraction from dog heart.

Dieter, Gross, and Marmé (for references, see Reference 14) demonstrated that membrane vesicles from various vegetables such as corn (*Zea mays*), squash (*Cucurbita pepo*), oats (*Avena sativa*), and mustard (*Sinapsis alba*), in the presence of ATP and Mg²⁺, accumulate Ca²⁺ against a concentration gradient. Although no conclusive answer can be given yet to the question of the origin of these membrane vesicles, the ATP-dependent Ca²⁺ transport can be stimulated by purified calmodulin either from plants or from animals and is inhibited by fluphenazine. The vesicles from corn cells

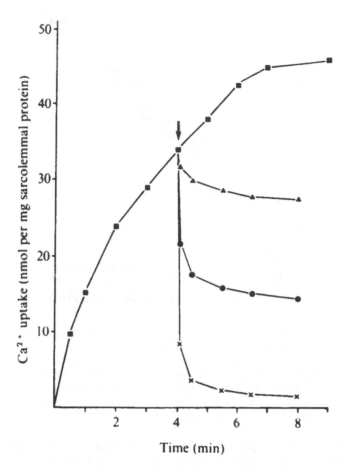

FIGURE 4. Na$^+$-induced releases of the Ca^{2+} accumulated by dog sarcolemma vesicles in the presence of ATP. ■, No additions; ▲, ●, and x, 10, 20 and 30 mM NaCl, respectively. (From Caroni, P. and Carafoli, E., *Nature*, 283, 765, 1980. With permission.)

have a Ca^{2+}-ATPase that can be purified by calmodulin-affinity chromatography. These results are important in showing that the Ca^{2+} pump system may also be present in the plasma membrane of cells from the vegetable kingdom.

The series of findings mentioned above suggested that the Ca^{2+} pump from plasma membrane, which for 12 years had appeared to pertain only to the red cell, had in fact a much wider distribution, being also present in cells as diverse as the squid giant axon and the dog heart muscle and hence that the ATP-dependent Ca^{2+} transport could be a general property of the plasma membrane of eukaryotic cells, as is already known for the Na$^+$ pump.

The previously mentioned findings, together with the increasing knowledge of the role of Ca^{2+} in cell metabolism, had led to the search for Ca^{2+}-pumping ATPases in the plasma membrane of a wide variety of cells.

C. How to Identify a Ca^{2+}-Transporting System with the Ca^{2+} Pump from Plasma Membrane

An unambiguous way to identify a Ca^{2+}-pumping ATPase in a plasma membrane would be to demonstrate the transport of Ca^{2+} with high affinity, stoichiometrically coupled to the hydrolysis of ATP within an ample range of ATP concentration in a pure preparation of plasma membrane. Ideally, this identification should be accom-

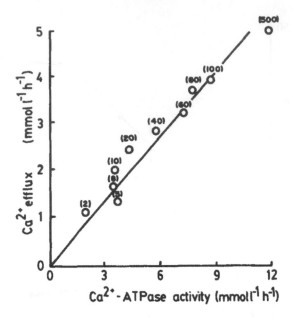

FIGURE 5.   The efflux of Ca²⁺ as a function of Ca²⁺-ATPase activity from human red cell ghosts. Figures in brackets are intracellular ATP concentration in μmol/ℓ. The slope of line that fits the points cannot be used as an estimate of stoichiometry since Ca²⁺ efflux was measured at higher Ca²⁺ concentration than Ca²⁺-ATPase activity. The data were taken from Figures 2 and 3, Reference 15.

plished by the isolation and purification to homogeneity of the Ca²⁺ pump and reconstitution in a lipid bilayer with the recovery of all the properties of the system in its original state. The work of Muallem and Karlish[15] on the regulation of the Ca²⁺ pump by ATP can be used to show the demonstration of stoichiometrical coupling between ATP hydrolysis and Ca²⁺ transport. The authors measured the rate of Ca²⁺ transport and Ca²⁺ dependent ATP hydrolysis at ATP concentrations between 1 and 1000 μM in resealed ghosts from human red cells where it was already known that Ca²⁺ is transported with high affinity. The plot of Ca²⁺ efflux vs. ATPase activity in Figure 5 can be fitted by a straight line, showing that the ratio Ca²⁺ efflux/Ca²⁺-ATPase activity remains constant over a wide range of ATP concentrations, a result that gives a strong argument in favor of the involvement of ATP hydrolysis in the transport of Ca²⁺ across plasma membrane of red cells. For many years, the red cell was the only one in which this sort of approach could be applied, mainly because of the simple structure of the cell and the availability of good techniques for the isolation and handling of their plasma membrane. This is one of the reasons why most of the present knowledge on the system comes from studies performed on either intact cells, isolated membranes, or purified enzyme from human red cells.

By contrast, obtaining pure preparations of plasma membrane from cell containing subcellular organelles is a difficult task, because membranes from the organelles could contaminate the preparation. This is particularly important, because organelles like mitochondria, endoplasmic reticulum, and sarcoplasmic reticulum possess calcium-transporting systems and/or hydrolyze ATP in a Ca²⁺-dependent fashion. Furthermore, as has been mentioned, plasma membrane preparations are capable of accumulating Ca²⁺ by a Na⁺-Ca²⁺ exchange mechanism which may operate in parallel with the Ca²⁺ pump and hence, the involvement of this mechanism must be discarded before attributing active transport of Ca²⁺ to the operation of a Ca²⁺-ATPase.

This, together with the lack of specific inhibitors of the Ca^{2+} pump, has made the search for Ca^{2+}-pumping ATPase in plasma membranes a difficult task. The discovery of the effects of calmodulin and its antagonists on the Ca^{2+} pump from plasma membranes has been of much help in this respect.

Knowledge of the properties of the Ca^{2+} pump, mostly gained from studies in human red cells, had provided guidelines for the identification of a Ca^{2+} pump in systems that do not allow the ideal criteria of identification to be met. Apart from its localization in the plasma membrane, among the properties that characterize a Ca^{2+}-transport system as a plasma membrane Ca^{2+} pump are

1. Under optimal assay conditions (low Mg^{2+}, calmodulin, pH = 7.4), the K_{c_a} for Ca^{2+} activation is 10^{-6} M or less.
2. The system depends on Mg^{2+} at micromolar concentrations. These concentrations are generally lower than those required by other ATPases activated by divalent cations that may be also present in plasma membranes.
3. It demonstrates specificity for ATP as the substrate.
4. There is Ca^{2+}-dependent formation of an acid-stable phosphoenzyme of rapid turnover and M_r 130,000 to 150,000. In red cells, the Km for the phosphorylation reaction is less than 10 μM ATP and the K_{c_a} is 10^{-6} M or less.
5. The system binds calmodulin, which increases the apparent affinity for Ca^{2+} and the turnover. Removal or blockage of endogenous calmodulin decreases the activity. When deprived of calmodulin, the Ca^{2+}-ATPase can be retained by affinity chromatography in a column of Sepharose-bound calmodulin.
6. In the presence of Mg^{2+} and K^+, orthovanadate inhibits the enzyme by combining with high affinity at a site on the internal surface of the membrane. The effect is antagonized by Ca^{2+} from the external surface of the membrane.
7. The activating effects of Mg^{2+}, Ca^{2+}, and ATP are exerted from the inner surface of the cell membrane.
8. The system shows specific immunoreactivity and complete lack of any immunological similarity with other cation transport ATPases.

The observer could look for these properties of the Ca^{2+} pump in either the Ca^{2+} transport or the ATP hydrolysis. Depending on the preparation, it could be easier to test one activity than the other. As was already mentioned for the giant axon, there are cells that do not provide one with enough material to measure Ca^{2+}-ATPase activity, but allow one to measure Ca^{2+} fluxes with high precision. Hence, in cells of this sort, it is advisable to study the characteristics of the fluxes of Ca^{2+} that depend on ATP. The opposite is also true for plasma membranes which do not recover spontaneously the low passive permeability to Ca^{2+} they have in the intact cell. If this is the case, the easiest way would be to characterize the system by its ability to hydrolyze ATP.

II. SOME PROPERTIES OF THE Ca^{2+} PUMP FROM VARIOUS CELL TYPES

Within the next paragraphs, some of the properties of the activities of the Ca^{2+} pump from the plasma membrane of different cells (except the erythrocyte) that can be found in the literature are summarized. The preparations listed are those which show at least one of the following properties: (1) they are pure plasma membrane; (2) Ca^{2+}-dependent activities are sensitive to calmodulin or calmodulin-antagonists, or (3) they show phosphoenzymes behaving as M_r 130,000 to 150,000 proteins in SDS-polyacrylamide gel electrophoresis.

Following Penniston's idea,[16] the cells have been classified as: circulating cells; excitable cells; tissue cells; and other cells.

Unless indicated, enzymic activities and Ca^{2+} fluxes are given in μmol/mg protein/min, and Km; K_i or K_{Ca} are in μM.

A. Circulating Cells

Lymphocyte — Lichtman et al.[17] Plasma membrane vesicles from human lymphocytes in the presence of ATP take up Ca^{2+} with Jm = 0.0024, K_{Ca} = 1, and Km = 80. ADP, AMP, GTP, UTP, ITP, or CTP do not substitute for ATP as the energy source. The vesicles hydrolyze ATP with Vm = 0.025, K_{Ca} = 0.6, and Km = 90. After treating the vesicles with EGTA, both Ca^{2+}-transport and ATP-hydrolysis are stimulated twofold by calmodulin.

Lymphocyte — Sarkadi et al.[18] Human lymphocytes loaded with Ca^{2+} by a short exposure to the A23187 Ca^{2+} ionophore show an ATP-dependent efflux of 40 μmol Ca^{2+}/min/l cells. Inside-out plasma membrane vesicles prepared from the lymphocytes take up Ca^{2+} from media containing ATP at a rate of 0.0012. The uptake is stimulated by calmodulin and by K^+.

Monocyte — Scully et al.[19] Human monocyte membrane vesicles loaded with oxalate in the presence of ATP accumulate Ca^{2+} with Jm = 0.004 and K_{Ca} = 0.53 and hydrolyze ATP with Vm = 0.018 and K_{Ca} = 0.6. No other nucleotide triphosphate substitutes for ATP as the substrate. Washing with EGTA lowers both transport and ATPase activities which are restored to their original levels by calmodulin.

Neutrophil — Ochs and Reed.[20] Plasma membrane vesicles from guinea pig neutrophil leukocytes incubated in the presence of ATP and oxalate accumulate Ca^{2+} with Jm = 0.017 and K_{Ca} = 0.164. GTP, ITP, CTP, UTP, ADP, or AMP do not replace ATP as the substrate. The vesicles hydrolyze ATP with Vm = 0.017 and K_{Ca} = 0.44. 50 μM trifluoperazine inhibits Ca^{2+} transport by near 90%. Exogenous calmodulin does not increase Jm nor decrease the K_{Ca} of the uptake by the vesicles.

B. Excitable Cells

Squid axon — DiPolo,[21] DiPolo and Beaugé.[22] Dialyzed squid axons show an ATP-dependent uphill extrusion of Ca^{2+} of 150 fmol/cm²/sec and K_{Ca} = 0.18 which persists in the absence of external Ca^{2+}, Na^+, or Mg^{2+}. In the absence of external Na^+, the Km for ATP during Ca^{2+} efflux is 30.

Optic nerve — Condrescu et al.[23] A membrane fraction enriched in axolemma from optic nerves of the squid show a Ca^{2+}-ATPase activity with high affinity for Ca^{2+} (K_{Ca} = 0.12 μM) and sensitivity to vanadate. Calmodulin stimulates the ATPase activity, and the stimulation is abolished by trifluoperazine.

Brain — Sorensen and Mahler.[24] Synaptic membrane preparations from rat cerebral cortex show Ca^{2+}-ATPase activity with Vm = 0.087 and K_{Ca} = 0.38. Calmodulin increases Vm to 0.161 and lowers K_{Ca} to 0.18. The activity is inhibited by VO_3^- with high affinity ($Ki \simeq 10$) and by phenothiazines and lanthanum. Na^+ or K^+ inhibit the Ca^{2+}-ATPase from EGTA-treated membranes and are without effect in the presence of calmodulin.

Brain — Hakim et al.[7] Synaptic plasma membranes from rat brain exhibit Ca^{2+}-ATPase activity with Vm = 0.058. K_{Ca} in the presence of calmodulin is 0.4. Calmodulin increases Vm and lowers K_{Ca}. The enzyme can be purified by calmodulin affinity chromatography up to a specific activity of 1.8 ± 0.2. Phosphorylation of the purified enzyme renders a phosphoenzyme of M_r 138,000. The purified enzyme cross-reacts with antibodies raised against the Ca^{2+} pump from human red cells. The authors suggest that the 100,000 M_r phosphoenzyme from rat brain reported by Robinson (see Chapter 4, Section I.B) may be a product of either proteolysis or maltreatment of the plasma membrane ATPase described here.

Neurohypophysis — Conigrave et al.[25] Plasma membrane vesicles from ox neurohypophysis nerve endings show ATP-dependent Ca^{2+} uptake with Jm = 1.66 and Ca^{2+}-ATPase activity. Both activities have the same dependence on Ca^{2+} and calmodulin concentration. Ca^{2+} uptake is inhibited by trifluoperazine.

Heart sarcolemma — Kuwayama and Kanazawa.[28] Vesicles of cardiac sarcolemma are isolated from canine ventricular muscle by centrifugation in a density gradient. In The vesicles also show ATP-dependent Ca^{2+} uptake of Jm = 0.031 and Na^+-Ca^{2+} exchange. The K_{Ca} = 0.3 is the same for ATP hydrolysis as well as for Ca^{2+} uptake. After washing with EGTA, the rates of ATP hydrolysis and Ca^{2+} uptake lower, and the K_{Ca} increases up to 11 μM. This result can be duplicated using trifluoperazine. The Ca^{2+}-ATPase activity is maximum at pH 7.3, is slightly increased by 160 mM K, and is inhibited in parallel with Ca^{2+} transport by increasing concentrations of vanadate (Ki = 0.5). The Ca^{2+}-ATPase is purified by means of a calmodulin-affinity chromatography column. One major protein band of M_r 150,000 is obtained which shows Ca^{2+}-ATPase activity and can be phosphorylated by ATP in a Ca^{2+}-dependent reaction. The purified ATPase can be reconstituted in asolectin liposomes where it pumps Ca^{2+}. The purified ATPase cross-reacts with an antibody raised against the Ca^{2+} pump from red cells.

Heart sarcolemma — Kuwayama and Kanayawa.[28] Vesicles of cardiac sarcolemma are isolated from canine ventricular muscle by centrifugation in a density gradient. In the presence of ATP, the vesicles pump Ca^{2+} which is released by Na^+ in the suspending medium, indicating that the Ca^{2+} pump was in the same membrane as the Na^+:Ca^{2+} antiporter, a recognized plasma membrane enzyme. Ca^{2+} uptake is stimulated by calmodulin. The activation by calmodulin is abolished by trifluoperazine.

Intestine muscle — Wibo et al.[29] A microsomal fraction from the longitudinal smooth muscle of guinea pig ileum made up of various subcellular entities takes up Ca^{2+}. In the presence of ATP, the uptake is stimulated by oxalate. The fraction insensitive to oxalate represents the uptake by the plasmalemmal vesicles. The rate of ATP-dependent Ca^{2+} uptake by plasmalemmal vesicles is near 0.020, is stimulated 30 to 40% by calmodulin, is inhibited by VO_3^- with Ki $\simeq$ 4, and is insensitive to K^+.

Stomach muscle — Wuytack et al.[30,31] A microsomal fraction from the smooth muscle of pig stomach shows an ATP-dependent Ca^{2+} uptake of approximately 0.005 and Ca^{2+}-ATPase activity of 0.011. Solubilization of the microsomes with deoxycolate followed by removal of the detergent by dialysis results in an increase in the specific activity by 18-fold. During dialysis, vesicles were reconstituted which were capable of ATP-dependent Ca^{2+} transport. Calmodulin-Sepharose affinity chromatography allows the isolation of a purified preparation of Ca^{2+}-ATPase with specific activity 338 × that of the crude microsomes. SDS gel electrophoresis of the purified ATPase shows a prominent band at M_r 140,000 that is phosphorylated by ATP.

Artery muscle — Wuytack and Casteels,[32] Wuytack et al.[33] A microsomal fraction of porcine coronary artery shows a Ca^{2+} uptake with Jm = 0.015 and K_{Ca} = 0.34 and Ca^{2+}-ATPase activity with Km = 0.020 and K_{Ca} = 1.17. Calmodulin increases Ca^{2+} uptake by a factor of 1.44, whereas the Ca^{2+}-ATPase is stimulated fivefold by calmodulin.

C. Tissue Cells

Intestine — Mellans and Popovich.[34] Basolateral plasma membrane vesicles from rat small intestine accumulates Ca^{2+} in the presence of ATP with Jm = 0.0012 and K_{Ca} = 0.03. Ca^{2+} uptake is inhibited by vanadate and stimulated by calmodulin which increases both the maximal transport rate and the calcium affinity of the transport mechanism.

Intestine — Ghijsen and Van Os,[35] De Jonge et al.,[36] Ghijsen et al.[37] Basolateral plasma membrane vesicles from rat duodenum epithelial cells accumulate Ca^{2+} with Jm

= 0.005 and K_{Ca} = 0.2 in the presence of ATP. Neither β-γ-(CH$_2$)-ATP, ADP, nor PNPP substitute for ATP. The vesicles show Ca²⁺-ATPase activity with Vm = 0.019 and K_{Ca} = 0.25. After washing with 5 mM EGTA, calmodulin increases the initial rate of Ca²⁺ uptake by more than 50%. In the presence of 1 μM Ca²⁺, ATP phosphorylates a protein with an apparent molecular weight of 115,000. Phosphorylation was strongly inhibited by phenothiazines.

Kidney — Gmaj et al.,[38] Gmaj et al.[39] Basolateral plasma membrane vesicles from the outer kidney cortex of rats, which in addition contain a Na⁺-Ca²⁺ exchange system, accumulate Ca²⁺ in an ATP-dependent way with Jm = 0.001 and K_{Ca} = 0.5. The membranes also show a Ca²⁺-ATPase activity with Vm = 0.080 and K_{Ca} = 0.68. This activity is stimulated by calmodulin and inhibited by vanadate with high affinity (Ki = 0.8).

Kidney — De Smedt et al.[40] Basolateral membrane preparations from dog kidney transport Ca²⁺ at a rate of 0.003. The membranes bind radioactive phosphorus when incubated with (γ-³²P)ATP in the presence of Ca²⁺. Two hydroxylamine-sensitive phosphoproteins are formed which, in SDS-polyacrylamide gel electrophoresis, migrate as M_r 130,000 and 100,000 proteins. The larger phosphoprotein comigrates with the Ca²⁺-ATPase from red cells.

Shell gland — Coty and McConkey.[41] A membrane fraction from hen oviduct shell gland exhibits a Ca²⁺-ATPase with Km = 0.150 and K_{Ca} = 0.4. The Km during ATP hydrolysis is 140 and neither ADP, AMP, GTP, nor ITP replace ATP as the substrate. Calmodulin is without effect on the rate of ATP hydrolysis, but trifluoperazine inhibits the ATPase activity. Incubation of ATP in the presence of Ca²⁺ leads to the formation of a phosphoprotein which in SDS polyacrylamide gel electrophoresis migrates as a protein of M_r 105,000.

Pancreas — Pershadsingh et al.,[42] Kotagal et al.[43] Plasma membranes from pancreatic islet cells have a Ca²⁺-ATPase with Vm = 0.054 and K_{Ca} = 0.09. As a function of ATP, the Ca²⁺-ATPase activity reveals two saturable components: a high-affinity component with Km = 2.1 and a low-affinity component with Km = 70. Membrane vesicles formed during fractionation accumulate Ca²⁺ in the presence of ATP at a rate of 0.010. Both Ca²⁺ transport and Ca²⁺-ATPase are activated by calmodulin and Ca²⁺-ATPase activity is inhibited by phenothiazines in a dose-response fashion.

Bone — Shen et al.[44] A plasma-membrane-enriched fraction from cultured bone cells exhibits Ca²⁺-ATPase activity with Vm = 0.013 and K_{Ca} = 0.28. The ATP activation curve is biphasic. The activity is slightly stimulated by 20 mM K⁺. Calmodulin lowers K_{Ca} to 0.05. Trifluoperazine lowers the Ca²⁺-ATPase activity which can be restored to its control value by exogenous calmodulin.

Adipocyte — Pershadsingh and McDonald,[45] Pershandsingh et al.[46] Rat adipocyte plasma membranes contain a Ca²⁺-ATPase with K_{Ca} = 0.14 and Vm = 0.097. Two activity components are distinguishable as a function of ATP. The high-affinity component has a Km = 1.1 and Vm = 0.006, and the data are insufficient for a reliable estimation of the Km of the low-affinity component. In the presence of ATP, the membrane transports Ca²⁺ with Jm = 0.03 and K_{Ca} = 0.19. Transport was stimulated approximately threefold by calmodulin at low Ca²⁺ concentration.

Corn coleoptile — Gross and Marmé,[47] Dieter and Marmé.[48] A plasma-membrane-enriched microsomal fraction from corn coleoptiles (*Zea mays* L.) in the presence of ATP shows Ca²⁺ uptake with Jm = 0.0003 and Ca²⁺-ATPase activity with Vm = 0.049. After partial purification by calmodulin-affinity chromatography, the microsomal ATPase is enhanced about twofold by calmodulin. Fluphenazine inhibits the stimulation by calmodulin.

D. Other Cells

Ehrlich ascites — Spitzer et al.[49] Plasma membrane vesicles of Ehrlich ascites carcinoma cells exhibit Ca²⁺-ATPase activity with Vm = 0.033. In the presence of ATP, the vesicles accumulate Ca²⁺ at a rate of 0.0003 with K_{Ca} = 0.25. The Km for ATP during Ca²⁺ uptake is 44. Calmodulin has no effect on Ca²⁺ transport. Phosphorylation of the vesicles with (γ-³²P)ATP reveals a Ca²⁺-dependent acylphosphate phosphoprotein of Mr 135,000.

Ehrlich ascites — Klaven et al.[50] In the presence of ATP, a vesicle preparation from Ehrlich ascites tumor cells highly enriched in plasma membranes accumulates Ca²⁺ at a rate of 0.12 and K_{Ca} = 0.031. Two Ca²⁺ stimulated ATPase activities were detected: one has a low apparent affinity (K_{Ca} = 136) and the other a high apparent affinity (K_{Ca} = 0.10) for Ca²⁺. The high-affinity ATPase appears to be related with Ca²⁺ transport. Trifluoperazine inhibits Ca²⁺ uptake and calmodulin activates Ca²⁺ uptake and the high-affinity Ca²⁺-ATPase activity.

Sperm — Breitbart et al.[51] Vesicles of purified membranes from ram spermatozoa accumulate Ca²⁺ in the presence of ATP with Jm = 0.0005, K_{Ca} = 2.5, and Km = 45. The vesicles hydrolyze ATP with Vm = 0.025 and K_{Ca} and Km values of 4.5 and 110 respectively. Ca²⁺ uptake and Ca²⁺-ATPase activity are both inhibited by quercetin, with similar effectiveness. The vesicles are 15-fold enriched in (Na + K)-ATPase and contain less than 4% of the cytochrome C oxidase specific activity found in whole cell homogenates.

E. Conclusion

The first conclusion that comes out from the description of the properties of the activities related to the Ca²⁺ pump made before is that the system is amply distributed among the different cell types. Cells from tissues and organisms as diverse as human blood, rat kidney, hen shell gland, and plants exhibit these activities, giving support to the conclusion that the Ca²⁺ pump system is present in most (if not all) plasma membrane from eukaryotic cells.

The activities of the Ca²⁺ pump cannot be compared among the different preparations without difficulties. This is mainly because the nature of the membrane preparations and the conditions in which they were assayed could have been not identical. This is particularly important when measuring Ca²⁺ fluxes, since in most cases it is difficult to assess the size of the compartments between which Ca²⁺ is transported. Hence, care has to be taken in comparing the magnitude of Ca²⁺ fluxes among cell membrane preparations as those cited above, because it does not permit one to reach any valid conclusion. Measurements of Ca²⁺-ATPase activities are less influenced by these factors and — provided they have been performed in the presence of calmodulin (or in membranes not totally rid of calmodulin) and nonlimiting concentrations of ATP and cofactors — to a certain extent allow a comparison of the total capacity of the Ca²⁺ pump among cell systems. Table 1 summarizes the data on Ca²⁺-ATPase activity and K_{Ca}. From all the cell membranes listed except the red cell, the mean value of Ca²⁺-ATPase activity can be calculated to be 0.050 μmol/mg protein/min, the lower value being 0.010 μmol/mg protein/min reported for dog heart sarcolemma and the higher value 0.150 μmol/mg protein/min for hen shell gland. The main conclusion to be drawn from these values is that plasma membranes do not hydrolyze ATP in a Ca²⁺-dependent fashion at high rate and that this rate does not vary largely among the different cell types. Ca²⁺-ATPase activity in sarcoplasmic reticulum can reach values of 2 μmol/mg protein/min[54] that is forty times that reported here as a mean value for plasma membranes. Since the turnover of the two Ca²⁺ transporting systems are similar (see Chapter 5, III.C), this would mean that plasma membranes are much less rich than sarcoplasmic reticulum in Ca²⁺ pump units. As judged by the number of calmodulin binding sites, there are 2000 copies of the Ca²⁺ pump per human red blood cell. Assuming that there are 10¹³ red cells and 10 mg of membrane protein per liter of red cells and a M_r 150,000

Table 1

MAXIMUM VELOCITY AND K_{Ca} OF THE Ca²⁺-ATPase IN DIFFERENT CELL TYPES

Cell type	Vm (μmol/mg protein/min)	K_{Ca} (μM)	Ref.
Human lymphocytes	0.025	0.6	17
Human monocytes	0.018	0.6	19
Human neutrophils	0.017	0.44	20
Rat brain	0.087	0.38	24
	0.058	0.40	7
Dog heart sarcolemma	0.10	0.3	28
Pig stomach smooth muscle	0.011	—	30
Porcine artery smooth muscle	0.020	—	32
Rat duodenum epithelial cells	0.019	0.25	35
Rat kidney	0.080	0.68	39
Hen shell gland	0.150	0.40	41
Rat pancreas	0.054	0.09	42
Bone cells	0.013	0.28	44
Rat adipocyte	0.097	0.14	45
Corn coleoptile	0.049	—	48
Ehrlich ascites	0.033	—	49
Ram spermatozoa	0.025	4.5	51
Human red blood cells			
(+ calmodulin)	0.065	—	52
(− calmodulin)	0.014	—	52
Purified ATPase from human red blood cells			
(+ calmodulin)	16.2	1.5	53
(− calmodulin)	2.1	20.0	53

for the Ca²⁺ pump, it can be calculated that in a red cell, the protein of the Ca²⁺ pump represents 0.05% of the membrane protein. The second property of the Ca²⁺ pump mentioned previously concerns its variation in activity among different cell types. It is illustrative to compare it with the Na⁺ pump, a transport system that contrasts sharply with the Ca²⁺ pump from plasma membranes in that respect. Bader et al.[55] reported the (Na⁺ + K⁺)-ATPase activity from human erythrocyte membranes and from electroplax electric cells as 0.007 and 3.17 μmol/mg protein/min, respectively; that is a range in specific activity more than 450-fold. This value is thirty times higher than the 15-fold range in the specific activity of the Ca²⁺ pump from plasma membranes reported here. The relatively low and similar activity of the Ca²⁺ pump of plasma membranes from different sources may well be related to the role in the control of cytosolic Ca²⁺ over long periods that has been assigned to the Ca²⁺ pump in Chapter 1.

REFERENCES

1. Dunham, E. T. and Glynn, I. M., Adenosinetriphosphatase activity and the active movements of alkali metal ions, *J. Physiol.,* 156, 274, 1961.
2. Schatzmann, H. J. and Vincenzi, F. F., Calcium movements across the membrane of human red cells, *J. Physiol.,* 201, 369, 1969.
3. Baker, P. J., Transport and metabolism on calcium ions in nerve, in *Prog. Biophys. Mol. Biol.,* Butler, J. A. V. and Noble, D., Eds., 24, 177, 1972.
4. Robinson, J. D., (Ca⁺ Mg)-stimulated ATPase activity of a rat brain microsomal preparation, *Arch. Biochem. Biophys.,* 176, 366, 1976.

5. Duncan, C. J., Properties of the Ca^{2+}-ATPase activity of mammalian synaptic membrane preparations, *J. Neurochem.*, 27, 1277, 1976.

6. Robinson, J. D., Calcium-stimulated phosphorylation of a brain (Ca^+ Mg)-ATPase preparation, *FEBS Lett.*, 87, 261, 1978.

7. Hakim, G., Itano, T., Verman, A. K., and Penniston, J. T., Purification of the Ca^{2+} and Mg^{2+} requiring ATPase from rat brain synaptic plasma membrane, *Biochem. J.*, 207, 225, 1982.

8. DiPolo, R., Reguena, J., Brinkley, F. J., Mullins, L. J., Scarpa, A., and Tiffert, T., Ionized calcium concentration in squid axons, *J. Gen. Physiol.*, 67, 433, 1976.

9. DiPolo, R., Ca pump driven by ATP in squid axons, *Nature*, 274, 390, 1978.

10. Cittadini, A. and van Rossum, Properties of the calcium-extruding mechanisms of liver cells, *J. Physiol.*, 281, 29, 1978.

11. Gmaj, P., Murer, H., and Kine, R., Calcium ion transport across plasma membranes isolated from rat kidney cortex, *Biochem. J.*, 178, 549, 1979.

12. Caroni, P. and Carafoli, E., An ATP-dependent Ca^{2+}-pumping system in dog heart sarcolemma, *Nature*, 283, 765, 1980.

13. Morcos, N. C. and Drummond, G. I., (Ca^{2+} + Mg^{2+})-ATPase in enriched sarcolemma from dog heart, *Biochim. Biophys. Acta*, 598, 27, 1980.

14. Marmé, D. and Dieter, P., Role of Ca^{2+} and calmodulin in plants, in *Calcium and Cell Function*, Vol. 4, Cheung, W. Y., Ed., Academic Press, New York, 1983, chap. 7.

15. Muallem, S. and Karlish, S. J. D., Is the red cell calcium pump regulated by ATP?, *Nature*, 277, 238, 1979.

16. Penniston, J. T., Plasma membrane Ca^{2+} ATPases as active Ca^{2+} pumps, in *Calcium and Cell Function*, Vol. 4, Cheung, W. Y., Ed., Academic Press, New York, 1983, chap. 3.

17. Lichtman, A. H., Segel, G. B., and Lichtman, M. A., Calcium transport and calcium-ATPase activity in human lymphocyte plasma membrane vesicles, *Biol. Chem.*, 256, 12, 1981.

18. Sarkadi, B., Enyedi, A., Szas, I., and Gárdos, G., Active calcium transport and calcium-dependent membrane phosphorylation in human peripheral blood lymphocytes, *Cell Calcium*, 3, 163, 1982.

19. Scully, S. P., Segel, G. B., and Lichtman, M. A., Plasma membrane vesicles prepared from unadhered monocytes: characterization of calcium transport and the calcium ATPase, *Cell Calcium*, 3, 515, 1982.

20. Ochs, D. L. and Reed, P. W., ATP-dependent calcium transport in plasma membrane vesicles from neutrophil leukocytes, *J. Biol. Chem.*, 258, 10116, 1983.

21. DiPolo, R., Ca pump driven by ATP in squid axons, *Nature*, 274, 390, 1978.

22. DiPolo, R. and Beaugé, L., Physiological role of ATP-driven calcium pump in squid axon, *Nature*, 278, 271, 1979.

23. Condrescu, M., Asses, L., and DiPolo, R., Partial purification and characterization of the (Ca^{2+} + Mg^{2+})-ATPase from squid optic nerve plasma membrane, *Biochim. Biophys. Acta*, 769, 281, 1984.

24. Sorensen, R. G. and Mahler, H. R., Calcium-stimulated adenosine triphosphatases in synaptic membranes, *J. Neurochem.*, 37, 6, 1981.

25. Conigrave, A. D., Treiman, M., Saermark, T., Thorn, N. A., Stimulation by calmodulin of Ca^{2+} uptake and (Ca^{2+}-Mg^{2+})-ATPase activity in membrane fractions from ox neurohypophyses, *Cell Calcium*, 2, 125, 1981.

26. Caroni, P. and Carafoli, E., The Ca^{2+}-pumping ATPase of heart sarcolemma, *J. Biol. Chem.*, 256, 3263, 1981.

27. Caroni, P., Zurini, M., Clark, A., and Carafoli, E., Further characterization and reconstitution of the purified Ca^{2+}-pumping ATPase of heart sarcolemma, *J. Biol. Chem.*, 258, 7305, 1983.

28. Kuwayama, H. and Kanazawa, T., Purification of cardiac sarcolemma vesicles: high sodium pump content and ATP-dependent, calmodulin-activated calcium uptake, *J. Biochem.*, 91, 1419, 1982.

29. Wibo, M., Morel, N., and Godfraind, T., Differentiation of Ca^{2+} pumps linked to plasma membrane and endoplasmic reticulum in the microsomal fraction from intestinal smooth muscle, *Biochim. Biophys. Acta*, 649, 651, 1981.

30. Wuytack, F., De Schutter, G., and Casteels, R., Partial purification of (Ca^{2+} + Mg^{2+})-dependent ATPase from pig smooth muscle and reconstitution of an ATP-dependent Ca^{2+}-transport system, *Biochem. J.*, 198, 265, 1981.

31. Wuytack, F., De Schutter, G., and Casteels, R., Purification of (Ca^{2+} + Mg^{2+})-ATPase from smooth muscle by calmodulin affinity chromatography, *FEBS Lett.*, 129, 297, 1981.

32. Wuytack, F. and Casteels, R., Demonstration of a (Ca^{2+} + Mg^{2+})-ATPase activity probably related to Ca^{2+} transport in the microsomal fraction of porcine coronary artery smooth muscle, *Biochim. Biophys. Acta*, 595, 257, 1980.

33. Wuytack, F., De Schutter, G., and Casteels, R., The effect of calmodulin on the active calcium-ion transport and (Ca^{2+} + Mg^{2+})-dependent ATPase in microsomal fractions of smooth muscle compared with that in erythrocytes and cardiac muscle, *Biochem. J.*, 190, 827, 1980.

34. Mellans, H. N. and Popovich, J. E., Calmodulin-regulated, ATP-driven calcium transport by basi-lateral membranes of rat small intestine. *J. Biol. Chem.,* 256, 9932, 1981.
35. Ghijsen, W. E. J. M. and Van Os, C. H., Ca-stimulated ATPase in brush border and basilateral membranes of rat duodenum with high affinity sites for Ca ions, *Nature,* 279, 802, 1979.
36. De Jonge, H. R., Ghijsen, W. E. J. M., and Van Os, C. H., Phosphorylated intermediates of Ca²⁺-ATPase and alkaline phosphatase in plasma membranes from rat duodenal epithelium, *Biochim. Biophys. Acta,* 647, 140, 1981.
37. Ghijsen, W. E. J. M., De Jong, M. D., and Van Os, C. H., ATP-dependent calcium transport and its correlation with Ca²⁺-ATPase activity in basilateral plasma membranes of rat duodenum, *Biochim. Biophys. Acta,* 689, 327, 1982.
38. Gmaj, P., Murer, H., and Kinne, R., Calcium ion transport across plasma membranes isolated from rat kidney cortex, *Biochem. J.,* 178, 549, 1979.
39. Gmaj, P., Murer, H. and Carafoli, E., Localization and properties of a high-affinity (Ca²⁺ + Mg²⁺)-ATPase in isolated kidney cortex plasma membranes, *FEBS Lett.,* 144, 226, 1982.
40. De Smedt, H., Parys, J. B., Borghgraef, R., and Wuytack, F., Phosphorylated intermediates of (Ca²⁺ + Mg²⁺)-ATPase and alkaline phosphatase in renal plasma membranes, *Biochim. Biophys. Acta,* 728, 409, 1983.
41. Coty, W. A. and McConkey, C., Jr., A high-affinity calcium-stimulated ATPase activity in the hen oviduct shell gland, *Arch. Biochem. Biophys.,* 219, 444, 1982.
42. Pershadsingh, H. A., McDaniel, M. L., Landt, M., Bry, C. G., Lacy, P. E., and McDonald, J. M., Ca²⁺-activated ATPase and ATP-dependent calmodulin-stimulated Ca²⁺ transport in islet cell plasma membrane, *Nature,* 288, 492, 1980.
43. Kotagal, N., Patker, C., Landt, M., McDonald, J., Colca, J., Lacy, P., and McDaniel, M., Regu-lation of pancreatic islet-cell plasma membrane (Ca²⁺ + Mg²⁺)-ATPase by calmodulin, *FEBS Lett.,* 137, 249, 1982.
44. Shen, V., Kohler, G., and Peck, W. A., A high affinity, calmodulin-responsive (Ca²⁺ - Mg²⁺)-ATPase in isolated bone cells, *Biochim. Biophys. Acta,* 727, 230, 1983.
45. Pershadsingh, H. A. and McDonald, J. M., A high affinity calcium-stimulated magnesium-depend-ent adenosine triphosphatase in rat adipocyte plasma membranes, *J. Biol. Chem.,* 255, 4087, 1980.
46. Pershadsingh, H. A., Landt, M., and McDonald, J. M., Calmodulin-sensitive ATP-dependent Ca²⁺ transport across adipocyte plasma membranes, *Biol. Chem.,* 255, 8983, 1980.
47. Gross, J. and Marmé, D., ATP-dependent Ca²⁺ uptake into plant membrane vesicles, *Proc. Natl. Acad. Sci. U.S.A.,* 75, 1232, 1978.
48. Dieter, P. and Marmé, D., A calmodulin-dependent, microsomal ATPase from corn (*Zea mays* L.), *FEBS Lett.,* 125, 245, 1981.
49. Spitzer, B., Bohmer, F. D., and Grosse, R., Identification of Ca²⁺-pump related phosphoprotein in plasma membrane vesicles of Ehrlich ascites carcinoma cells. *Biochim. Biophys. Acta,* 728, 59, 1983.
50. Klaven, N. B., Pershadsingh, H. A., Henius, G. V., Laris, P. C., Long, J. W., and McDonald, J. M., A high-affinity calmodulin-sensitive (Ca²⁺ + Mg²⁺)-ATPase and associated calcium-transport pumping in the Ehrlich ascites tumor cell plasma membrane, *Arch. Biochem. Biophys.,* 226, 618, 1983.
51. Breitbart, H., Stern, B., and Rubinstein, S., Calcium transport and Ca²⁺-ATPase activity in ram spermatozoa plasma membrane vesicles, *Biochim. Biophys. Acta,* 728, 349, 1983.
52. Scharff, O., Kinetics of calmodulin-dependent (Ca²⁺ + Mg²⁺)-ATPase in plasma membranes and sol-ubilized membranes from erythrocytes, *Arch. Biochem. Biophys.,* 209, 72, 1981.
53. Gietzen, K. and Koland, J., Large-scale isolation of human erythrocyte Ca²⁺-transport ATPase, *Biochem. J.,* 207, 155, 1982.
54. de Meis, L., The sarcoplasmic reticulum, in *Transport in the Life Sciences,* Bittar, E. E., Ed., Vol. 2, John Wiley & Sons, New York, 1981.
55. Bader, H., Post, R. L., and Bond, G. H., Comparison of sources of phosphorylated intermediate in transport ATPase, *Biochim. Biophys. Acta,* 150, 41, 1968.

Chapter 5

ISOLATION AND PURIFICATION OF THE Ca²⁺ PUMP

P. J. Garrahan

I. THE MAIN DIFFICULTIES AND THE FIRST ATTEMPTS

The isolation and purification of the plasma membrane Ca²⁺ pump is essential for detailed structural studies and for the unambiguous identification of those Ca²⁺-dependent enzymatic activities of membrane preparations which are catalyzed by the Ca²⁺ pump.

In the sarcoplasmic reticulum of skeletal muscle, up to 70% of the membrane protein is the Ca²⁺-pumping ATPase.[1] In some plasma membranes, up to 50% of the membrane protein can be the $(Na^+ + K^+)$-pumping ATPase.[2] Given this abundance, the main technical obstacle for the purification of these ATPases is their separation from the other components of the membrane without irreversible loss of their function. This problem was overcome at the beginning of the 1970s with the development of procedures for solubilizing membrane components with dilute aqueous solutions of detergents. Once this was achieved, the relative abundance of the transport ATPases as compared to other membrane components made it easy to apply conventional techniques for purifying these systems (for references, see References 1 and 2).

In contrast with what happens with the sarcoplasmic reticulum and Na⁺ pump ATPases (as has been mentioned in Chapter 4, Section II.E), there does not seem to be a plasma membrane that is "naturally enriched" in Ca²⁺-ATPase. Hence, one of the main problems researchers had to face when trying to purify the Ca²⁺-ATPase of plasma membranes was that only a minute fraction of the total membrane protein pertains to this system. This makes it very difficult to apply conventional protein fractionation techniques to membrane solubilizates. An additional difficulty which soon became apparent was the unstability of the solubilized Ca²⁺-ATPase which leads to its quick and irreversible inactivation after solubilization.[3]

The first successful partial purification of the Ca²⁺-ATPase of plasma membranes was reported in length in 1977 by Wolf and co-workers.[3] These authors used dilute aqueous solutions of Triton X-100® to solubilize red blood cell membranes and showed that the quick inactivation of the Ca²⁺-ATPase could be avoided if mixed micelles of the nonionic detergent Tween-20® and crude phosphatidyl choline were present in the solubilization media. By chromatography on Sepharose CL-6B, a substantial purification of the solubilized enzyme was achieved, specific activity increasing from 0.02 to 3.2 mol/mg protein/min. The enzyme retained the kinetic properties of the native enzyme. The authors performed all the procedures in the presence of protease inhibitors since they showed that the enzyme was very sensitive to endogenous proteolysis. SDS gel electrophoresis of the partially purified enzyme revealed the presence of three protein bands of apparent M_r 145,000, 115,000 and 105,000. The component of highest molecular weight was phosphorylated by ATP.

Although the purification procedure of Wolf et al. has now been superseded by techniques based on affinity chromatography, the pioneering work of this group set general guidelines which are still valid for obtaining soluble preparations which are apt for purification.

The functional competence of Wolf's enzyme was confirmed in 1978 by Haaker and Racker[4] who showed that Ca²⁺-ATPase, isolated from pig erythrocyte membranes and reconstituted into asolectin (crude soybean phospholipids) liposomes, is able to sustain an uptake of Ca^{2+} that is coupled with a ratio of 1:1 to the hydrolysis of ATP. This was the first successful functional reconstitution of the Ca^{2+} pump of plasma membranes.

In 1978, Peterson et al.[5] reported another successful attempt to solubilize the Ca^{2+}-ATPase. They showed that after solubilization of red blood cell membranes with Triton X-100®, Ca^{2+}-ATPase activity could be regained if the detergent is removed and the solubilized ATPase is incorporated into phosphatidyl serine or oleic acid liposomes. The authors tried to purify the enzyme by isoelectric focusing of the solubilizate obtaining 14-fold increase in specific activity, a value which is much lower than that reported by Wolf et al.[3]

In 1979, Gietzen et al.[6] solubilized red blood cell membranes with deoxycholate and by means of the "cholate dialysis" procedure developed by Meissner and Fleischer[7] for sarcoplasmic reticulum and achieved a successful reconstitution of Ca^{2+}-pumping activity.

II. THE USE OF CALMODULIN AFFINITY CHROMATOGRAPHY TO PURIFY THE Ca²⁺-ATPase

Affinity chromatography is in many cases the procedure of choice for separating components which, like the Ca^{2+}-ATPase, are present in minute amounts in complex mixtures.

Calmodulin affinity chromatography can be used for the purification of a calmodulin-dependent enzyme if the following two criteria are satisfied. (1) The enzyme has to bind with high affinity to calmodulin. This does not follow from its calmodulin dependence since, as we have discussed in Chapter 3, activation by calmodulin can be mediated by a calmodulin-dependent phosphokinase. (2) The enzyme has to be the only calmodulin-binding protein in the mixture, since calmodulin-Sepharose columns will retain all substances that bind calmodulin.

By the time the conditions for successful solubilization of the Ca^{2+}-ATPase were being established, it was known that in the presence of Ca^{2+}, calmodulin interacted reversibly and with high affinity with the ATPase (see Chapter 10), and that, at least in red blood cell membranes, the Ca^{2+}-ATPase is the only component binding calmodulin with high affinity that remains after washing the membranes with low-ionic-strength solutions. This made calmodulin an attractive candidate as a ligand for affinity chromatography to purify the ATPase by means of calmodulin-Sepharose conjugates. The rationale of this procedure is as follows:

1. Pure calmodulin is covalently coupled to CNBr-Sepharose and a calmodulin-Sepharose column is equilibrated with a buffer containing Ca^{2+} and phospholipids.
2. A solubilized preparation of membranes stabilized by lipids in media containing Ca^{2+} is passed through the column and then the column is thoroughly washed with Ca^{2+}-containing buffers. As a consequence of this, all membrane components will be eluted except the Ca^{2+}-ATPase that remains bound to the immobilized calmodulin.
3. Since calmodulin only binds to the ATPase in the presence of Ca^{2+}, the enzyme will elute from the column when Ca^{2+} is removed with a Ca^{2+} chelator like EDTA.

Purification of the plasma membrane Ca^{2+}-ATPase using calmodulin affinity chromatography was first reported in 1979 by Niggli et al.[8] These authors used red blood cell membranes solubilized with Triton X-100® and stabilized with phosphatidyl serine. Most of the membrane proteins including the Mg^{2+}-dependent ATPase activity was eluted with Ca^{2+} containing buffers. When 5 mM EDTA replaced Ca^{2+} in the

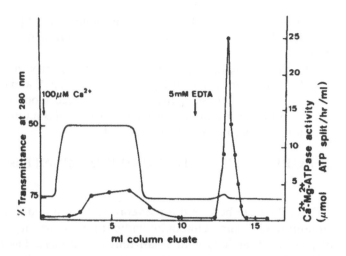

FIGURE 1.   Light transmittance at 280 nm (—) and Ca²⁺-ATPase activity (•—•) of the eluate after affinity chromatography of Triton X-100®-solubilized red blood cell membranes on a Sepharose-43-calmodulin column. Where indicated buffers containing either Ca²⁺ or EDTA were used for elution. (From Niggli, V., Penniston, J. T., and Carafoli, E., *J. Biol. Chem.*, 254, 9955, 1979. With permission.)

elution buffer, a small symmetric peak was eluted containing the Ca^{2+}-dependent ATPase and about 0.09% of the protein of the starting material (see Figure 1). The specific activity of the Ca^{2+}-ATPase in this peak was 3.8 as compared to 0.09 $\mu mol/mg$ protein/min in the solubilizate. In unidimensional SDS gel electrophoresis, the peak containing the ATPase showed a major band of M_r 125,000 and a minor band of M_r 205,000. Both bands were phosphorylated by ATP in the presence of Ca^{2+} and Mg^{2+}. A property of the purified enzyme (which was puzzling at that time) was that it could no longer be activated by calmodulin (see Chapter 10).

Five months after the paper by Niggli et al. was published, Gietzen et al.[9] reported the second successful purification of red blood cell membrane Ca^{2+}-ATPase using calmodulin-affinity chromatography. These workers employed deoxycholate for solubilization and phosphatidyl choline to stabilize the enzyme. The results obtained were similar to those of Niggli et al. except that: (1) the specific activity of the purified material was 3 to 4 times higher, (2) the activity of the purified enzyme could be increased about nine times by exogenous calmodulin, and (3) the apparent M_r of the enzyme was 145,000. It is now known that the different reactivity to calmodulin between the two preparations was caused by the different phospholipids employed to stabilize the enzyme, rather than by an intrinsic difference between the two enzymes; this is discussed in detail in Chapter 10. The higher apparent M_r of this enzyme with respect to that of Niggli et al. probably resulted from the fact that Gietzen et al. protected their preparation against endogenous proteolytic activities.

After the initial reports, Graf et al.[10] showed that the specific activity of the purified enzyme could be improved, freeing it of minor impurities by extensive washings while bound to the calmodulin-Sepharose column. These studies also cancelled the initial discrepancies about the molecular weight of the main protein band. The functional competence of the purified enzyme was proved by showing that it could be reconstituted into phospholipid vesicles and support ATP-dependent active Ca^{2+} uptake into these vesicles.[11-14] In 1982, a rapid large-scale procedure for purification of the Ca^{2+}-ATPase from red blood cell membranes based on calmodulin affinity chromatography was proposed by Gietzen and Kolandt.[15] This method allows to process in a day material from up to 5 ℓ of packed red blood cells. It yields about 5 mg of purified ATPase per liter of packed cells which corresponds to 0.2% of the total membrane protein of the starting material.

Calmodulin-affinity chromatography has also been used to purify the Ca²⁺-ATPase of heart sarcolemma,[16,17] of rat brain synaptic plasma membranes,[18] and of pig antrum smooth muscle.[19] Up to this date, these — together with the red blood cell membrane enzyme — are the only ATPases unambiguously identified as plasma membrane Ca²⁺-pumping ATPases, that have been isolated and purified.

III. PROPERTIES OF THE PURIFIED Ca²⁺-ATPase

A. Stability

The purified Ca²⁺-ATPase can be stored for several weeks at liquid nitrogen temperature in media containing Ca²⁺ and calmodulin without loss in activity.[15] At 0°C, the enzyme is unstable, losing its activity with a half-time of 1.2 days. The half-time can be increased to 6 days if calmodulin and Ca²⁺ are present in the media.[15,21]

B. Molecular Weight and Composition

The protein moiety of the ATPase is a single polypeptide chain whose apparent M_r estimated by SDS gel electrophoresis is 140,000. This value is similar to the apparent M_r given in Chapter 8 for the membrane protein that is phosphorylated by ATP in the presence of Ca²⁺. No carbohydrates are detectable in the purified enzyme.[10,14] The enzyme contains about 7 mol of tightly bound phosphate per mole of enzyme.[14] It is not known if this phosphate belongs to tightly bound phospholipids or if it is phosphate covalently linked to amino acid residues. Using radioimmunoassay, Gietzen and Kolandt[15] estimated that the purified preparation contains about 1 mol of EGTA-dissociable calmodulin per 400 mol of ATPase. Hence the enzyme can be considered to be almost devoid of calmodulin. The purified Ca²⁺-ATPase shows a strong tendency to aggregate forming dimers and higher order oligomers that fail to dissociate in SDS, 7 M urea, or mercaptoethanol.[10] It is likely that these aggregates are the higher molecular weight components that often appear in SDS gel electrophoresis of the purified enzyme. The molecular weight of the functional Ca²⁺-ATPase has been studied by Cavieres[20] measuring radiation inactivation of the Ca²⁺-ATPase activity of red blood cell membranes. An average target size of 251 kdaltons was obtained. This is close to twice the apparent M_r of the polypeptide chain of the Ca²⁺-ATPase and suggests that the active pump is a dimer of the polypeptide.

The amino acid composition of the ATPase has been determined (see References 10 and 14 and Table 1). Calculations based on this composition indicate that the ATPase is only moderately hydrophobic. This suggests that a significant part of the molecule is not embedded in the lipid bilayer, but is exposed to the aqueous solutions that bathe the membrane.

As is described in detail in Chapter 10, a large polypeptide (M_r about 30,000) can be cleaved from the enzyme by controlled proteolysis with trypsin without loss in catalytic and transport properties but with complete disappearance of calmodulin sensitivity. Hence an important fraction of the protein moiety of the Ca²⁺-ATPase is not directly involved in catalysis or in transport, but seems to pertain to a regulatory domain of the molecule. If the M_r of the putative regulatory domain is subtracted from the M_r of the whole enzyme, the resulting M_r becomes close to that of the main polypeptide chain of the Ca²⁺-ATPase of sarcoplasmic reticulum and of the (Na⁺, K⁺)-ATPase of plasma membranes which in both cases is about 100,000.[2,3]

Since the polypeptide that is removed by proteolysis is accessible to attack by a large hydrophilic enzyme like trypsin, it is reasonable to think that it pertains to a region of

Table 1
AMINO ACID COMPOSITION
OF Ca^{2+}-ATPase

Amino acid	mol/mol[a]
Lys	84
His	22
Arg	56
Asx	131
Thr	73
Ser	85
Glx	134
Pro	62
Gly	95
Ala	81
Cys	20
Val	97
Met	29
Ile	83
Leu	115
Tyr	26
Phe	49
Trp	7

[a] Based on a molecular weight of 138,000.

Taken from Graf, E. et al., *Biochemistry*, 21, 4511, 1982. With permission.

the ATPase molecule that is hydrophilic and exposed to the solutions that bathe the membrane. This is consistent with the comments we made above concerning the amino acid composition of the ATPase.

C. Kinetic Properties

The specific activity of the purest samples of Ca^{2+}-ATPase is about 15 to 20 μmol/ mg protein/min, and the percentage of the total activity that is not dependent on Ca^{2+} is below 1%. Assuming these samples are 100% pure and the active enzyme is monomeric, the maximum measured specific activities would correspond to a turnover number of about 50 sec^{-1}. This value is close to that which can be calculated in intact red blood cells from the ratio of maximum Ca^{2+}-ATPase activity (0.1 μmol/mg protein/ min) to steady-state phosphoenzyme level (1 pmol/mg protein).

The specific activity of the Ca^{2+}-ATPase is close to that reported for purified preparations of sarcoplasmic reticulum Ca^{2+}-ATPase[1] and of plasma membrane (Na$^+$, K$^+$)-ATPase.[2] This strongly suggests that the turnover rate of all cation-transport ATPases is on the same order of magnitude.

The purified Ca^{2+}-ATPase shares most of the kinetic properties of the enzyme in unfractioned membranes. These are discussed in detail in the relevant sections of this book and include: (1) the ability to couple the hydrolysis of ATP to the active transport of Ca^{2+};[11-14] (2) high-affinity activation by Ca^{2+} and modulation of this affinity by calmodulin, the lipid environment, and limited proteolysis; (3) high specificity for ATP as the substrate[10] and a biphasic response to the nucleotide with high- and low-affinity components;[21] (4) Ca^{2+}-dependent phosphorylation by ATP and (ATP + Mg^{2+})-dependent dephosphorylation;[14,21] and (5) activation of the red cell enzyme by Na$^+$ and K$^+$.[10]

D. Reconstitution of the Purified Enzyme

The availability of purified preparations of Ca²⁺-ATPase has allowed detailed studies on the reconstitution of active transport in artificial lipid membranes. These studies, apart from demonstrating the functional competence of the purified ATPase, have yielded useful information on lipid-dependence of the Ca²⁺ pump (see Chapter 10), on the stoichiometry of active Ca²⁺ transport, and on the charge-balance during this process (see Chapter 7).

E. Immunological Reactivity

Verma et al.[22] have raised antibodies directed against the purified Ca²⁺-ATPase from red blood cell membranes. The antibodies inhibit ATPase activity and active Ca²⁺ transport and have no effect on calmodulin binding.

In competitive radioimmunoassay tests of immunological cross reactivity, the red blood cell ATPase showed a consistent pattern of immunological similarity to the Ca²⁺ ATPases of other plasma membranes such as those of rat and dog red blood cells, rat corpus luteum, and rat brain synaptosomes.[22] In sharp contrast, purified Ca²⁺-ATPase from rabbit sarcoplasmic reticulum failed to show any immunological similarity to the red blood cell enzyme. This constitutes strong evidence that the plasma membrane and the sarcoplasmic reticulum Ca²⁺ pumps are different molecular entities.

REFERENCES

1. de Meis, L., The sarcoplasmic reticulum. Transport and energy transduction, in *Transport in the Life Sciences*, Vol. 2, Bittar, E. E., Ed., John Wiley & Sons, New York, 1981.
2. Joergensen, P. L., Purification of the (Na⁺ K⁺)-ATPase, Active site determination and criteria of purity, *Ann. N.Y. Acad. Sci.*, 242, 36, 1974.
3. Wolf, H. U., Diekvoss, G., and Lichtner, R., Purification and properties of high-affinity Ca²⁺-ATPase of human erythrocyte membranes, *Acta Biol. Med. Ger.*, 36, 847, 1977.
4. Haaker, H. and Racker, E., Purification and reconstitution of the Ca²⁺-ATPase from plasma membranes of pig erythrocytes, *J. Biol. Chem.*, 254, 6589, 1979.
5. Peterson, S. W., Ronner, P., and Carafoli, E., Partial purification and reconstitution of the (Ca²⁺-Mg²⁺)-ATPase of erythrocyte membranes, *Arch. Biochim. Biophys.*, 186, 202, 1978.
6. Gietzen, K., Seiler, S., Fleischer, S., and Wolf, H. U., Reconstitution of the Ca²⁺-transport system of human erythrocytes, *Biochem. J.*, 188, 47, 1980.
7. Meissner, G. and Fleischer, S., Dissociation and reconstitution of functional sarcoplasmic reticulum vesicles, *J. Biol. Chem.*, 249, 302, 1974.
8. Niggli, V., Penniston, J. T., and Carafoli, E., Purification of the (Ca²⁺ - Mg²⁺)-ATPase from human erythrocyte membranes using a calmodulin affinity column, *J. Biol. Chem.*, 254, 9955, 1979.
9. Gietzen, K., Tejcka, M., and Wolf, H. U., Calmodulin affinity chromatography yields a functionally purified erythrocyte (Ca²⁺ + Mg²⁺)-dependent adenosine triphosphatase, *Biochem. J.*, 189, 81, 1980.
10. Graf, E., Verma, A. K., Gorski, J. P., Lopaschuk, G., Niggli, V., Zurini, M., Carafoli, E., and Penniston, J. T., Molecular properties of the calcium-pumping ATPase from human erythrocytes, *Biochemistry*, 21, 4511, 1982.
11. Niggli, V., Adunyah, E. S., Penniston, J. T., and Carafoli, E., Purified (Ca²⁺ - Mg²⁺)-ATPase of the erythrocyte membrane: reconstitution and effect of calmodulin and phospholipids, *J. Biol. Chem.*, 256, 395, 1981.
12. Niggli, V., Sigel, E., and Carafoli, E., The purified Ca²⁺ pump of human erythrocyte membranes catalyzes an electroneutral Ca²⁺ - H⁺ exchange in reconstituted liposomal systems, *J. Biol. Chem.*, 257, 2350, 1982.
13. Niggli, V., Adunyah, E. S., and Carafoli, E., Acidic phospholipids, unsaturated fatty acids and limited proteolysis mimic the effect of calmodulin on the purified erythrocyte Ca²⁺-ATPase, *J. Biol. Chem.*, 256, 8588, 1981.
14. Carafoli, E. and Zurini, M., The Ca²⁺-pumping ATPase of plasma membranes. Purification reconstitution and properties, *Biochim. Biophys. Acta*, 683, 279, 1982.

15. Gietzen, K. and Kolandt, J., Large scale isolation of human erythrocyte Ca^{2+}-transport ATPase, *Biochem. J.,* 207, 155, 1982.
16. Caroni, P., Zurini, M., and Clark, A., The calcium-pumping ATPase of heart sarcolemma, *Ann. N.Y. Acad. Sci.,* 402, 402, 1982.
17. Caroni, P. and Carafoli, E., The Ca^{2+}-pumping ATPase of heart sarcolemma. Characterization, calmodulin dependence and partial purification, *J. Biol. Chem.,* 256, 3263, 1981.
18. Hakim, G., Itano, T., Verma, A. K., and Penniston, J. T., Purification of the Ca^{2+} and Mg^{2+}-requiring ATPase from rat brain synaptic plasma membrane, *Biochem. J.,* 207, 225, 1982.
19. Wuytack, F., De Shutter, G., and Casteels, R., Purification of $(Ca^{2+} + Mg^{2+})$-ATPase from smooth muscle by calmodulin affinity chromatography, *FEBS Lett.,* 129, 297, 1981.
20. Cavieres, J. D., Calmodulin and the target size of the $(Ca^{2+} + Mg^{2+})$-ATPase of human red cell ghosts, *Biochim. Biophys. Acta,* 771, 241, 1984.
21. Stieger, J. and Luterbacher, S., Some properties of the purified $(Ca^{2+} + Mg^{2+})$-ATPase from human red blood cell membranes, *Biochim. Biophys. Acta,* 641, 270, 1981.
22. Verma, A. K., Gorski, J. P., and Penniston, J. T., Antibodies toward human erythrocyte Ca^{2+}-ATPase: effect on enzyme function and immunoreactivity of Ca^{2+}-ATPases from other sources, *Arch. Biochim. Biophys.,* 215, 345, 1982.

Chapter 6

TRANSPORT OF Ca²⁺ AND ATP HYDROLYSIS BY THE Ca²⁺ PUMP

A. F. Rega

I. TRANSPORT OF Ca²⁺

A. Introduction

As shown in Chapter 3, the transport of Ca²⁺ across plasma membranes has been demonstrated in a number of eukaryotic cells. To study kinetics of active Ca²⁺ transport through the plasma membrane in cells possessing intracellular organelles is a difficult task, because in the presence of these organelles that could transport Ca²⁺ and consume or produce ATP, it is almost impossible to keep constant known concentrations of Ca²⁺ and ATP in the cytosol through the experiment. Thus, most transport studies have been made in vesicular preparations derived from the plasma membrane rather than in intact cells. Many times these vesicular preparations are enriched — rather than pure — plasma membrane preparations that could be contaminated by intracellular membranes able to transport Ca²⁺. The only membrane component of the red blood cell is the plasma membrane. This, together with the fact that it was first demonstrated in it, are the main reasons that for a long time the only membrane preparations used to study the Ca²⁺ pump were those from red blood cells. The advantages of the red blood cell for transport studies have also been well appreciated during the characterization of the Na²⁺ pump. Most of the knowledge on properties such as the asymmetrical requirement of Na⁺ and K⁺, the stoichiometry of the transport reaction, the alternative ways of functioning, etc. of the Na⁺ pump comes from experiments made in red blood cells.

B. Preparations Used for Transport Studies

Four kinds of membrane preparations have shown to be suitable to study Ca²⁺ fluxes across the plasma membrane, i.e.: (1) intact red blood cells, (2) resealed ghosts from red blood cells, (3) inside-out vesicles (IOVs), (4) reconstituted liposomes, and (5) squid axons.

1. Intact Red Blood Cells

Measurements of Ca²⁺ efflux from intact red blood cells require the cells to be loaded with Ca²⁺. Since the passive permeability of the cell membrane to Ca²⁺ is very low (see Table 1 of Chapter 2), it has to be increased to facilitate the penetration of the cation into the cell interior. Two procedures have been used for loading the cells with Ca²⁺. Schatzmann[1] applied the PCMBS (p-chloromercurybenzene sulfonate) procedure[2] to human red blood cells. The cells suspended in an isotonic solution containing PCMBS and Ca²⁺ are stored in the cold overnight. They are then washed in the same solution containing cysteine and no PCMBS and incubated at 37°C in the cysteine medium for 1 hr. After the mercurial has been removed, the cells regain the low permeability to Ca²⁺ and upon incubation at 37°C with Pi, adenine, and inosine, the Ca²⁺ that has been retained in the cell is pumped out. Intact human red blood cells can be loaded with Ca²⁺ by means of the divalent-cation ionophore A23187.[3,4] For this purpose, the cells are incubated at 37°C in an isotonic solution containing the ionophore and Ca²⁺ during 2 min. They are then washed with an isotonic solution containing albumin. After the ionophore has been washed out, the passive permeability of the membrane to Ca²⁺ returns to its normal value, the cells remaining without significant change relative to

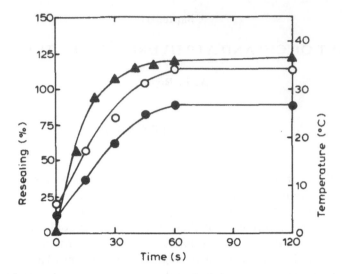

FIGURE 1.   Time course of the temperature (▲) and the resealing of a suspension of red blood cell ghosts to Ca^{2+} (●) and ATP (O) in a water-bath at 37°C. (From Kratje, R. B., Garrahan, P. J., and Rega, A. F., *Biochim. Biophys. Acta,* 731, 40, 1983. With permission.)

untreated cells in their content of Na^+, K^+, and ATP. Loading intact cells with Ca^{2+} allows one to measure Ca^{2+} fluxes in the presence of most of the cell components and it is recommended when one wishes to measure fluxes under conditions almost identical to those prevailing in intact cells. Furthermore, since the ionophore-induced increase in Ca^{2+} permeability in the cell population is uniform,[5] intracellular Ca^{2+} in A23187-treated red blood cells should be homogeneously distributed among the cells.

An elegant nondisrupted technique for loading calcium buffers into cells has been described by Tsien.[6] It consists basically in making a Ca^{2+} chelator (a member of the family of the tetracarboxylate dyes mentioned in Chapter 1) temporarily membrane-permeable by masking its four carboxylates with acetyl groups to form an acetoxymethyl tetraester which is hydrolyzed by acetylesterases inside the cells. This regenerates and traps the original chelator that is a derivative of EGTA in which benzene rings replace the methylene groups connecting N to O. The chelator-loaded cells are then incubated in a Ca^{2+} containing medium with sufficient Ca-ionophore A23187 to ensure that intracellular Ca^{2+} reaches equilibrium with external Ca^{2+}. The method has been demonstrated in human erythrocytes, lymphocytes, and mast cells.[6]

2. Resealed Ghosts

The experiment in Figure 1 shows that it suffices for red blood cell ghosts to reach 37°C to recover the normal low permeability to Ca^{2+} and ATP of the red blood cell membrane. This property allowed us to develop a procedure for loading red blood cells with Ca^{2+} by reversible hemolysis that can be applied as follows.[7] Red cells are lysed in water at 0°C, and then the tonicity of the suspending medium is restored to its original value by the addition of a concentrated salt solution containing CaEDTA buffer and phosphocreatine plus creatine phosphokinase to regenerate ATP from ADP to avoid depletion of energy stores. Sealing is accomplished in a very short time (about 10 sec) by immersing the flask containing the ghost suspension in a water bath at 60°C until the temperature of the suspension reaches 37°C. Short incubation time reduces to a minimum the depletion of Ca^{2+} from the resealed ghosts and makes reasonable the assumption that, provided the ghosts have not changed their volume, immediately after sealing the concentration of Ca^{2+} inside the ghosts is equal to that in the solution in which they were sealed. The concentration of Ca^{2+} in the sealing solution can be measured with accuracy by means of a Ca^{2+}-selective electrode which is more reliable than estimations derived from equilibrium binding constants. Resealed ghosts have the advantage over intact red blood cells in that they can be filled with calcium buffers and ATP-regenerating systems of the proper composition to maintain almost constant a given concentration of intracellular Ca^{2+} and the rate of Ca^{2+} efflux during about 10 min at 37°C. Resealed ghosts are also suitable for simultaneous measurement of Ca^{2+} efflux and ATP hydrolysis.[8]

3. Inside Out Vesicles

Isolated plasma membranes undergo spontaneous vesiculization. A number of the vesicles are inside out (IOV) so that, in them, the external surface of the membrane that faced the suspending medium in the cell will now face the interior of the vesicle, and the opposite will occur with the surface of the membrane that in the intact cell was in contact with the cytoplasm. In IOVs, therefore, Ca^{2+} is transported from the external medium to the interior of the vesicle and ATP and Mg^{2+} are required in the external medium. In Chapter 4, Section II, many examples of the use of plasma membrane vesicles to study Ca^{2+} transport can be found. Inside out vesicles from human red cell membranes have become very popular in the study of Ca^{2+} transport. The methods of Steck and Kant[9] and of Lew et al.[10] provide IOVs which pump Ca^{2+} actively. Both procedures are based on the observation that, in the absence of divalent cations, red blood cell membranes in an aqueous medium of low ionic strength and pH near 8.0 undergo spontaneous vesiculization. If the vesicles are then sumitted to a shearing stress by passing the vesicle suspension through a small caliber hypodermic needle, iontight vesicles mostly homogeneous in size are obtained. Of the vesicles, 30 to 50% are IOVs.

The main advantages of IOVs for Ca^{2+} transport studies is that they allow one to know and manipulate the actual concentration of Ca^{2+}, ATP, and all the ligands that modify the transport process from the cytoplasmic surface of the cell membrane and are suitable for the measurement of Ca^{2+} transport and ATP hydrolysis in the same preparation.[11] Vesicles are generally small in volume. This was estimated to be 7 to 14 μm^3 by Sarkadi et al. in vesicles from red blood cells.[11] As a consequence of this, Ca^{2+} accumulated in the interior of the vesicle reaches concentrations which can be high enough as to affect the rate of Ca^{2+} uptake. This difficulty can be overcome (Figure 2) by incubating the vesicles in the presence of oxalate which equilibrates rapidly between the intra- and extravesicular media and lowers the concentration of intravesicular Ca^{2+}.[12] Although it can be argued that vesiculization could lead to changes in the native state of the membrane or in the properties of the transport system, there is no doubt that IOVs permit more detailed kinetic experiments than is possible with right side out preparations.

4. Reconstituted Liposomes

Ca^{2+} transport by the Ca^{2+} pump reconstituted in liposomes has been demonstrated for the purified ATPase from red blood cells and heart sarcolemma. Measuring fluxes in reconstituted liposomes implies the isolation or, at least, extensive purifications of the Ca^{2+} pump. However, the validity of results obtained with a preparation which is devoid of most of the other protein components of the plasma membrane justify the use of the system. Two procedures have been described to incorporate the Ca^{2+}-ATPase into asolectin (crude soybean phospholipids). According to the freeze-thaw sonication procedure, purified ATPase in a buffer containing asolectin is frozen in liquid nitrogen, thawed at room temperature for 20 min, and sonicated for 1 min under nitrogen.[13]

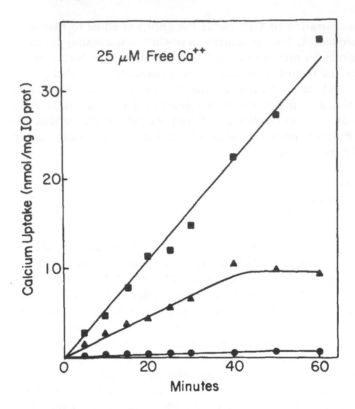

FIGURE 2.   ⁴⁵Ca uptake by IOVs from human red blood cell membranes as a function of time in control medium (●); in medium with 3 mM MgATP (▲) and in medium with 3 mM MgATP plus 5 mM sodium oxalate (■). (From Mollman, J. E. and Pleasure, D. E., *J. Biol. Chem.*, 255, 569, 1980. With permission.)

The procedure of cholate dialysis[14,15] consists in adding purified Ca²⁺-ATPase to asolectin dispersed with cholate in aqueous salt solutions. The mixture is submitted to vortex mixing and then dialyzed at 4°C against a neutral salt solution of almost physiological tonicity. The reconstituted liposomes obtained by any of the two procedures are not leaky to Ca²⁺ and show Ca²⁺ uptake coupled to the hydrolysis of ATP. Figure 3 shows the uptake of Ca²⁺ simultaneous with the hydrolysis of ATP by reconstituted liposomes of heart sarcolemma.

5. Squid Axons

All the studies on Ca²⁺ transport by the Ca²⁺ pump in squid axons that have been reported were peformed by DiPolo and Beaugé. They have published a clear and authoritative description of the general properties of the preparation and the procedure they follow to measure the fraction of the Ca²⁺ efflux that takes place through the Ca²⁺ pump.[16] According to them, the internal dialysis technique allows a precise control of the intracellular concentration of solutes during the course of the experiment. A glass or plastic capillary tube with a central porous region is steered through the axon. The porous region is permeable to solutes of molecular weight up to 1000, allowing an exchange of solutes between a fluid passed through the capillary and the axoplasm. To allow a better control of the intracellular Ca²⁺ concentration, the mitochondria are poisoned with CN⁻, oligomycin, and FCCP. In addition, EGTA is always present to overcome the buffering capacity of the axoplasm components. Intracellular medium must contain ATP and Mg to sustain active transport of Ca²⁺. To avoid any contami-

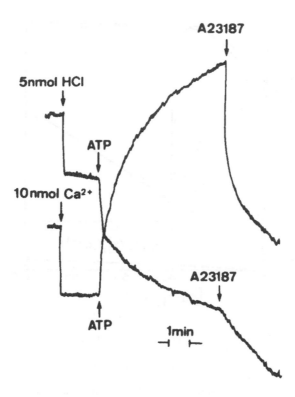

FIGURE 3. The time course of Ca²⁺ uptake and H⁺ production by ATP hydrolysis by purified Ca²⁺-ATPase from heart sarcolemma reconstituted into isolectin liposomes. Ca²⁺ pumping was initiated by the addition of ATP. Ca²⁺ uptake was followed by measuring the changes in Ca²⁺ concentration of the suspending medium by means of a Ca²⁺-selective electrode. ATP hydrolysis was estimated indirectly by the increase in H⁺ concentration. (From Caroni, P., Zurini, M., Clark, A., and Carafoli, E., *J. Biol. Chem.*, 258, 7305, 1983. With permission.)

nation with the Ca²⁺ efflux that takes place via the Na⁺:Ca²⁺ exchanger, measurements are performed in external media without Na⁺ and Ca²⁺. Under these conditions, most of the Ca²⁺ efflux from the axon takes place through the Ca²⁺ pump since the passive leaks are very small.[16]

C. Dependence on Ca²⁺ Concentration
1. Activation by Ca²⁺

Figure 4 shows the initial rate of the efflux of ⁴⁵Ca²⁺ from reconstituted ghosts of human red blood cells containing 100 mM choline, various amounts of Ca²⁺ and either 0 or 0.2 mM vanadate suspended in choline medium with and without 0.2 mM vanadate at 37°C. The ghosts were sealed in CaEGTA buffers whose Ca²⁺ content was measured with a Ca²⁺-selective electrode. The Ca²⁺ efflux in the presence of vanadate can be taken as the passive leak of Ca²⁺ from the cells (see Chapter 11, Section II.B). A double reciprocal plot of the total efflux (inset in Figure 4) gives a straight line showing that under the experimental conditions of the assay the total efflux of Ca²⁺ changes with intracellular Ca²⁺ along a single Michaelis curve with K_{Ca} 3.3 μM and a maximum rate of 21 mmol/l cells/hr or 0.035 μmol/mg protein/min (calculated assuming 10 mg membrane protein per ml of ghost). Since the lowest Ca²⁺ concentration tested was 1 μM, that is, 1/3 of K_{Ca}, the possibility of deviation from simple hyperbolic kinetics at very low Ca²⁺ concentration cannot be discarded.

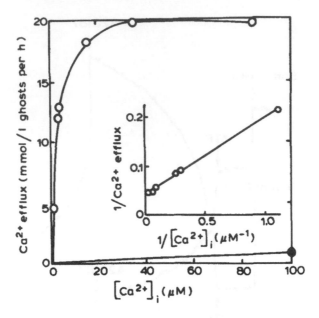

FIGURE 4.   ⁴⁵Ca²⁺ efflux from resealed ghosts containing 100 mM with (●) and without (○) 0, 2 mM vanadate as a function of the intracellular Ca²⁺ concentration. The ghosts with vanadate were incubated in solutions containing 0.2 mM vanadate. The inset represents a double reciprocal plot of the efflux in the absence of vanadate. (From Kratje, R. B., Garrahan, P. J., and Rega, A. F., *Biochim. Biophys. Acta*, 731, 40, 1983. With permission.)

Figure 5 shows results of an experiment designed to measure the dependence of Ca²⁺ concentration of active Ca²⁺ efflux from dialyzed squid axon. The K_{ca} is about 0.15 μM.

H. J. Schatzmann[17] reported that in resealed ghosts the shape of the efflux curve as a function of Ca²⁺ concentration is sigmoidal rather than hyperbolic. Such kinetic behavior could be taken as an indication that to be transported, two Ca²⁺ have to combine with the pump. However, Schatzmann was not fully in favor of this interpretation, arguing that during the experiments, ghosts containing low Ca²⁺ concentration were less tight to Ca²⁺ and some leakage of Ca²⁺ back into the cells might give an apparently reduced rate at low Ca²⁺ concentration which, of course, will lead to an S-shaped curve. In connection with the kinetics of activation by Ca²⁺ are the findings on the response of the Ca²⁺-ATPase to Ca²⁺ concentration in Section II.A of this chapter.

From the results of Figure 4, without taking into consideration the activation by K⁺ which is described in Chapter 9, Section II, the maximum rate of Ca²⁺ transport in intact red cells can be considered to be 20 mmol/l cells/hr. As the rate of Ca²⁺ influx into ATP-depleted red blood cells at 2 mM external Ca²⁺ given in Chapter 2, Section I.B is 0.030 mmol/l cells/hr, it can be concluded that the Ca²⁺ pump has to work at about 0.1% of its full capacity at most to maintain constant the level of cytosolic Ca²⁺ in circulating red cells. Although there might be circumstances that make the membrane more permeable to Ca²⁺ and the rate of Ca²⁺ influx larger than the value taken above, it seems that the capacity of the cell to pump Ca²⁺ out is in great excess.

2. Inhibition by Ca²⁺

It has been reported that concentrations of Ca²⁺ higher than those necessary for full activation of the Ca²⁺ pump are inhibitory so that the rate of Ca²⁺ transport as a func-

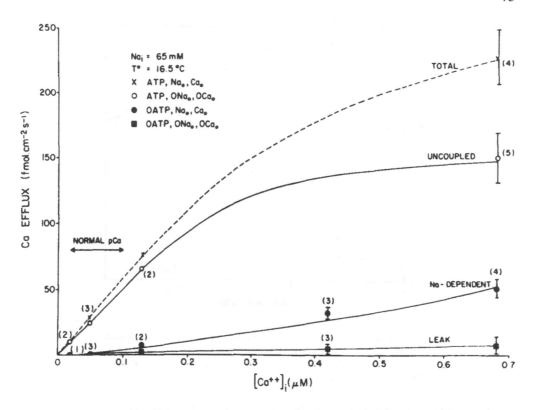

FIGURE 5. Effect of internal Ca^{2+} on the Ca^{2+} efflux components in dialyzed squid axons. Numbers above the experimental points represent the number of determinations. The physiological range of internal Ca^{2+} concentration is indicated by the horizontal arrow. (From DiPolo, R. and Beaugé, L., *Cell Calcium*, 1, 147, 1980.

tion of Ca^{2+} concentration gives a biphasic curve. Such behavior can be observed clearly in Figure 6 which shows the rate of Ca^{2+} uptake by IOVs from human red blood cells as a function of the concentration of Ca^{2+} in a suspending medium containing 3 mM ATP plus $MgCl_2$ and no EGTA. The concentration of Ca^{2+} was determined with arsenazo III. Each point represents the mean of eight determinations of the initial rate of Ca^{2+} influx. The influx rises at low calcium concentration reaching a maximum at about 50 μM Ca^{2+}. The activation can be represented by a Michaelis curve of K_{Ca} 3.4 μM and Vm near 0.010 μmol/mg protein/min showing no sign of sigmoidicity. As Ca^{2+} concentration rises above 50 μM, the rate of transport decreases, indicating that in IOVs, relatively high Ca^{2+} concentrations on the cytoplasmic side of the membrane are inhibitory. Inhibition of Ca^{2+} transport by Ca^{2+} in the 50 to 120 μM concentration range has not been observed by other authors.[18,19] However, in connection with this point, it is pertinent to inform the reader that inhibition of the Ca^{2+}-ATPase activity by relatively high Ca^{2+} concentration is now an accepted fact (see Section II.A.2 of this chapter).

R. B. Kratje (unpublished) has investigated in our laboratory the effects of Ca^{2+} applied on the external surface of cell membrane on the active transport of Ca^{2+} in resealed ghosts from human red blood cells. External Ca^{2+} inhibits active transport with low apparent affinity (Ki $\simeq$ 9 mM) along a rectangular hyperbola which tends to zero. The effect is independent of calmodulin, EGTA, or intracellular Ca^{2+}, is stimulated by alkaline pH, and, although with less apparent affinity, it can be reproduced by Mg^{2+}. It is not known yet whether the effect is exerted by combination of extracellular Ca^{2+} with the Ca^{2+} pump or not.

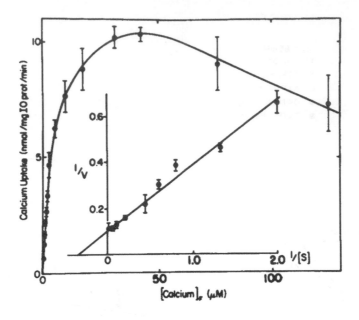

FIGURE 6.   Rate of Ca²⁺ uptake by IOVs from human red blood cells as a function of the concentration of Ca²⁺ in the suspending medium. (From Mollman, J. E. and Pleasure, D. E., *J. Biol. Chem.*, 255, 569, 1980. With permission.)

D. Substances and Treatments that Increase the Rate of Transport

Mild proteolysis, an acidic lipid environment, and calmodulin increase the apparent affinity for Ca²⁺ and the maximum rate of the active transport. Their effects are not additive and can be made apparent if the preparation used has been deprived of calmodulin. Furthermore, since neither calmodulin nor proteolytic enzymes can cross the membrane in intact cells or resealed ghosts, their effects on Ca²⁺ transport can only be tested in IOVs or reconstituted liposomes. The subject is treated in detail in Chapter 10.

It is well known that the maximum rate of Ca²⁺ transport in IOVs is increased by oxalate[12,20] as has been shown in Figure 2.

We have also tested the effect of 5 m*M* oxalate in IOVs from human red blood cells[21] and found that it increases 20-fold (from 0.005 to 0.1 μmol/mg protein/min) the maximum rate of Ca²⁺ transport whereas Ca²⁺-ATPase activity increases about 1.5-fold (from 0.0085 to 0.0145 mol/mg protein/min). The simplest explanation for this result is that oxalate retains intravesicular calcium which otherwise would have leaked out. However, the small but constant activation of the Ca²⁺-ATPase by oxalate we have observed is against the idea of attributing the activating effect of oxalate to retention of Ca²⁺ within the vesicle solely. A plausible explanation is based on the observation mentioned previously that external Ca²⁺ at millimolar concentration inhibits Ca²⁺ efflux from resealed ghosts. If this effect is also present in IOVs, accumulation of Ca²⁺ inside the vesicles will lead to progressive inhibition of Ca²⁺ transport. Under these conditions, inclusion of oxalate will lower the concentration of intravesicular Ca²⁺ and reverse inhibition of Ca²⁺ uptake and ATP hydrolysis.

Although no detailed studies have been made, it seems that reducing agents modify the maximum rate of transport. It has been reported that glutathione or dithiothreitol in the hemolysis solutions and the buffers in which IOVs from red blood cells are formed increase two- to tenfold the maximum rate of Ca²⁺ transport without changing the $K_{0.5}$ for Ca²⁺.[12] In human red blood cells, small anions like Cl⁻, HCO₃⁻ and H₂PO₄⁻

increase Ca^{2+} transport by a mechanism that does not involve direct interaction with the Ca^{2+} pump and which is described in the next section.

In intact cells the concentration of ATP is between 1 to 1.5 mM and that of free Mg^{2+} about 0.7 mM. These concentrations are well above those necessary for half-maximal activation of the Ca^{2+} pump by ATP or by Mg^{2+}; hence, changes within physiological ranges in the concentration of Mg^{2+} or of ATP will not affect significantly the activity of the pump. In contrast with this, the resting concentration of cytosolic Ca^{2+} is well below that needed for full stimulation. This makes it likely that of all the ligands that participate, only Ca^{2+} (and calmodulin, see Chapter 10) is involved in physiologically meaningful changes of the activity of the Ca^{2+} pump.

E. The Electrical Balance During Transport of Ca^{2+}

A question intimately related to the mechanism of the Ca^{2+} pump is whether the active transport of Ca^{2+} across the plasma membrane is coupled to the movement of other ions through the Ca^{2+} pump. A negative answer implies that the pump transports Ca^{2+} without compensating the two positive charges, the transport process being electrogenic. If Ca^{2+} transport is coupled to the transport of other ions, then the two positive charges could be compensated by cotransport or countertransport of the other ion species, the transport process being electroneutral. There are few ions that *a priori* could be considered to be partners of Ca^{2+} during active transport. Among them is Mg^{2+}, because it is essential for activation, and Na^+ or K^+, because they stimulate the pump activity. Cl^- and H^+ should be taken into account also, because they are always present during Ca^{2+} transport under physiological conditions. That the Ca^{2+} pump exchanges Ca^{2+} for Mg^{2+}, Na^+, or K^+ in an electroneutral process can be ruled out, because Mg^{2+} is needed at the cell interior and the monovalent cations activate Ca^{2+} transport from the internal surface of the cell membrane and are not essential for active transport of Ca^{2+}.[7]

Studies to answer the fundamental question posed above have mainly been performed on the Ca^{2+} pump from human red blood cells.

1. Electrogenic Transport

Gimble and his colleagues[22] made qualitative determinations of membrane potential by means of probes such as 3,3'-dipropylthiodicarbocyanine iodide, 1-anilino-8-naphthalenesulfonate, and an electron paramagnetic resonant, triphenylphosphonium, and found that during active Ca^{2+} uptake in media containing a nonpermeant anion like gluconate, IOVs from human red blood cells develop a membrane potential (positive inside). The fluorescent response of the probes to the membrane potential shows the same dependence on Ca^{2+}, ATP, and calmodulin as the Ca^{2+} transport suggesting that the response is a consequence of the active transport of Ca^{2+} into the vesicles. When phosphate is added to the gluconate medium that contain the vesicles, Ca^{2+} transport is stimulated in parallel to an ATP plus Ca^{2+}-dependent and calmodulin-stimulated uptake of phosphate by the IOVs.[23] The calcium ionophore A23187 releases the Ca^{2+} as well as the phosphate that has accumulated in the vesicles during Ca^{2+} transport. Furthermore, at concentrations in which they have little or no effect on Ca^{2+}-ATPase activity, inhibitors of the red blood cell anion transport system (band III) like DIDS (4,4-diisothiocyano-2,2-stilbenedisulfonic acid) applied on the membrane surface facing the interior of the vesicle inhibit with the same potency both Ca^{2+} and phosphate uptake.[23] Either sulfate, chloride, or acetate replace phosphate in stimulating Ca^{2+} uptake in gluconate medium.[24] The authors rationalized their results as follows: (1) Ca^{2+} transport through the Ca^{2+} pump leads to the development of a membrane potential positive inside the IOVs; (2) the membrane potential acts as the driving force for anion transport through band III; (3) the uptake of these anions causes the collapse of this membrane potential and, thus, promotes further accumulation of Ca^{2+} by the IOVs.[25]

Rossi and Schatzmann[26] used light scattering to measure volume changes of IOVs of high potassium permeability from human red cells. In a chloride medium, the vesicles swell during Ca^{2+} transport. When the impermeant gluconate anion is added in place of Cl^-, the vesicles shrink and the loss of K^+ from the vesicles increases. The volume changes are consistent with the assumption that in Cl^- media, one Ca^{2+} is cotransported with two Cl^- ions from the medium and in gluconate media, the Ca^{2+} that enters the vesicles is exchanged for two K^+. Rossi and Schatzmann[26] concluded that according to the conditions chosen, either Cl^- or K^+ movements through passive channels parallel to the pump provide electroneutrality of the overall ionic shift. This view agrees well with that of Gimble et al., the main conclusion being that the Ca^{2+} pump is electrogenic. A Ca^{2+}-ATPase preparation from sarcoplasmic reticulum isolated and reconstituted into liposomes containing phosphatidylcholine only, was shown to be electrogenic.[27]

2. Electroneutral Transport

Niggli et al.[28] studied Ca^{2+} transport by the purified Ca^{2+}-pumping ATPase from human red cells reconstituted in asolectin liposomes. This system seems ideal for investigating the charge balance during Ca^{2+} transport, because the reconstituted liposomes are tight to Ca^{2+} and have the Ca^{2+} pump as their only protein component. Thus neither band III nor other ion transport systems are expected to be present in the liposomes. The authors concluded that Ca^{2+} transport by the purified ATPase is electroneutral after finding that:

1. The potassium ionophore valinomycin does not stimulate Ca^{2+} transport by Ca^{2+}-ATPase that has been reconstituted and assayed in K^+ medium.
2. Membrane potentials (either negative or positive inside the vesicles) created by means of Na^+ or K^+ concentration gradients do not affect the initial rate of Ca^{2+} transport in the liposomes.
3. Under conditions in which DIDS inhibits 50% of Ca^{2+} uptake, it also inhibits 50% of Ca^{2+}-ATPase activity, and in media with K^+, the inhibition is not reversed by valinomycin.
4. No accumulation of tetraphenylboron (a lipophilic anion) takes place during Ca^{2+} transport.

Niggli et al.[28] also measured H^+ translocation during Ca^{2+} transport by suspending the reconstituted Ca^{2+} pump in a medium of low pH-buffering capacity. Under these conditions, addition of ATP leads to a rapid initial H^+ appearance in the suspending medium which slows down gradually after 30 sec. Addition of A23187 at this point stimulates the phenomenon which reaches a constant linear rate 60 to 70% lower than the initial rate in the absence of the ionophore. The H^+ produced in the presence of the ionophore represents H^+ due to ATP hydrolysis. The authors' interpretation is that the difference between the initial rate of H^+ production in the presence of ATP minus the H^+ due to ATP hydrolysis represents the initial rate of H^+ extrusion due to the uptake of Ca^{2+} by the Ca^{2+} pump. On this basis, the molar ratio H^+ extruded/ATP hydrolyzed is near 2 and on the assumption that one Ca^{2+} is transported per ATP hydrolyzed in reconstituted liposomes, 2 H^+ are extruded when one Ca^{2+} is taken up. After these findings, Niggli et al.[28] concluded that Ca^{2+} transport by the Ca^{2+} pump is electroneutral, and since the Ca^{2+} pump is the only protein component in the liposomes, the Ca^{2+} pump has to behave as a $Ca^{2+} - H^+$ (or $Ca^{2+} - OH^-$) exchanger.

This conclusion, based on experiments performed on the reconstituted system, is at variance with that based on experiments performed with red cell membranes containing

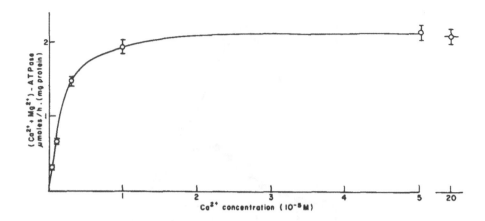

FIGURE 7. Ca²⁺-ATPase activity of isolated membranes from human red blood cells as a function of Ca² concentration. (From Schatzmann, H. J. and Roelofsen, B., in *Biochemistry of Membrane Transport*, Semenza, G. and Carafoli, E., Eds., Springer-Verlag, Berlin, 389, 1977. With permission.)

most of their protein constituents. The discrepancy is difficult to reconcile. More recently, Smallwood et al. proposed a new model for the erythrocyte Ca²⁺ pump in IOVs as a Ca²⁺:H⁺ exchanger.[29] Such an electroneutral exchange across the Ca²⁺ pump system is not harmonious with the membrane potential that, as mentioned above, the same group of authors detected in IOVs after the functioning of the pump.

II. ATP HYDROLYSIS

A. Dependence on Ca²⁺ Concentration

The experimental values of the apparent affinity for Ca²⁺ and the maximum ATPase activity measured at saturating Ca²⁺ are highly dependent on calmodulin, acidic phospholipids, partial proteolysis, EGTAs and other factors. In Chapters 7 and 10, the reader can find information on this particular, but important, aspect of the activation of the Ca²⁺-ATPase by Ca²⁺.

1. Activation by Ca²⁺

Schatzmann and Roelofsen[30] measured Ca²⁺-ATPase activity in the 10^{-6} to 10^{-5} M Ca²⁺ concentration range using a preparation of human red blood cell membranes. The membranes were prepared in the absence of Ca²⁺-chelating agents, because in this way, a homogeneous enzyme preparation, probably saturated with calmodulin, giving a single high-affinity response to Ca²⁺ is obtained.[1] The curve of Ca²⁺-ATPase activity vs. the calculated Ca²⁺ concentration in Ca-EGTA buffer is shown in Figure 7. The maximum rate of ATP hydrolysis is 0.033 μmol/min/mg protein and K_{Ca} about 2 μM. The curve shows a slight sigmoidicity, its Hill coefficient is 1.3, and differs from 1 in a statistically significant way. The validity of this result on the sigmoidicity of the curve and its implication on the cooperativity of the Ca²⁺-ATPase with respect to Ca²⁺ are commented below. The activation of the Ca²⁺-ATPase by Ca²⁺ has also been studied in plasma membrane preparations from cells as diverse as rat pancreatic islet cell,[31] guinea pig neutrophil leukocytes,[32] and ram spermatozoa plasma membrane.[33] For all these cells, the experimental points of the reciprocal of Ca²⁺-ATPase activity vs. the reciprocal of Ca²⁺ concentration up to 10 μM fall on straight lines, suggesting that in these membrane preparations, the activation of the Ca²⁺-ATPase by Ca²⁺ follows a Michaelis-Menten-like kinetics. Furthermore, Klaven et al.[34] made with much detail a Ca²⁺ curve of the Ca²⁺-ATPase activity from Ehrlich ascites tumor cells up to 1.73 μM Ca²⁺,

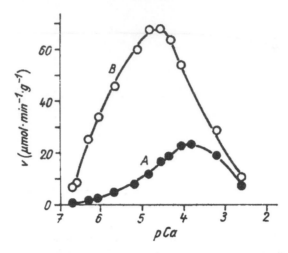

FIGURE 8.   Ca²⁺-ATPase activity of calmodulin-deficient (●) red blood cell membranes, and calmodulin-saturated (○) red cell membranes as a function of Ca²⁺ concentration. (From Scharff, O., *Acta Biol. Med. Germ.*, 40,
457, 1981. With permission.)

a concentration at which the activity reaches near saturation. The curve shows no sign
of sigmoidicity, since its Hill coefficient is 1.06 ± 0.11.

Apart from the small sign of cooperativity detected in the curve from Figure 6, it
has been reported that the Hill coefficient of the Ca²⁺ activation curves of the Ca²⁺-
ATPase from human erythrocytes increases from 1 to 1.5 to 2.0[35] and 3[36] in the presence of calmodulin. Does it mean that if combined with calmodulin, the Ca²⁺-ATPase
displays cooperative activation kinetics with respect to Ca²⁺? These results are discussed
in Chapter 7, Section I.C. In Chapter 4 can be found information and further comments on the values of Vm and K_{Ca} of Ca²⁺-ATPase activity from various types of cells.

2. Inhibition by Ca²⁺

Scharff[35] found that when the study of the dependence of the Ca²⁺-ATPase activity
on Ca²⁺ is extended to Ca²⁺ concentration in the millimolar range as in the experiment
from Figure 8, a new phenomenon becomes apparent. The curve relating Ca²⁺-ATPase
activity to Ca²⁺ concentration is biphasic, the activity reaches a maximum, and then
drops to almost zero as Ca²⁺ concentration rises. If calmodulin is present, the shape of
the curve remains the same, but its maximum is shifted towards lower Ca²⁺ concentration (Figure 8). Scharff's findings on the biphasic response of Ca²⁺-ATPase activity to
Ca²⁺ concentrations have been confirmed by others in human erythrocyte
membranes[37-39] and extended to Ca²⁺-ATPase from kidney cortex plasma membranes
and to purified Ca²⁺-ATPase from normal[40] and sickled human red blood cells reconstituted in phosphatidylcholine liposomes.[41] These findings, together with those on
Ca²⁺ transport shown in Section I.C of this chapter, seem to indicate that the biphasic
response to Ca²⁺ represents a general property of the Ca²⁺ pump of plasma membranes.

The concentration of Ca²⁺ that lowers to half the Ca²⁺-ATPase activity attained at
the maximum of the curve ranges within 10^{-4} to 10^{-3} M, and most authors agree that,
as is apparent at first glance in Figure 8, it is lowered by calmodulin. As a consequence
of this, it has been thought that the Ki for Ca²⁺ depends on calmodulin. This view has
not been shared by Schatzmann[42] on the grounds of data by Wuthrich[37] which show
that the inhibition is independent of calmodulin. It is not easy to solve this discrepancy

with the information which is available now, but if we accept that the biphasic curve is the sum of an activation curve whose K_{Ca} is lowered by calmodulin and an inhibition curve whose Ki is independent of calmodulin, in the presence of calmodulin, the maximum of the biphasic curve will be displaced towards low Ca^{2+} concentration, and, as in the case of Figure 8, the inhibition will now be apparent at lower Ca^{2+} concentrations as if calmodulin had increased the apparent affinity for Ca^{2+} during inhibition.

3. The Mechanism of the Inhibition

In interpreting the results of the inhibition of the Ca^{2+}-ATPase from plasma membranes, it has to be considered that the inhibitory species could be either Ca^{2+} or CaATP.

With respect to Ca^{2+}, the only experimental evidence showing that Ca^{2+} itself lowers the rate of the overall reaction catalyzed by the Ca^{2+} pump is the inhibition of Ca^{2+} transport in resealed ghosts from human red blood cells by Ca^{2+} applied externally, we have referred to previously. However, it is difficult to relate this phenomenon with the inhibition of the Ca^{2+}-ATPase described here, because the K_{Ca} for inhibition of the Ca^{2+}-ATPase is at least one order of magnitude less than the K_{Ca} for inhibition of Ca^{2+} transport by external Ca^{2+}.

We know by a series of experiments performed in our laboratory and described in Chapter 9, Section I.C, that Ca^{2+} inhibits Ca^{2+}-ATPase activity with Ki = 30 μM by displacing Mg^{2+} at a site from which Mg^{2+} activates the Ca^{2+}-ATPase. Furthermore, it has been reported that binding of Ca^{2+} to the Ca^{2+}-ATPase with Ki near 40 μM leads to inhibition of the dephosphorylation reaction,[43] an effect that, as judged by the Ki value, could have been caused by the displacement of Mg^{2+} mentioned before.

To test CaATP as the inhibitory ligand, Muallem and Karlish[38] measured ATP hydrolysis by human red blood cell membranes at 5, 50, 250, and 500 μM Ca^{2+} as a function of MgATP in media in which the calculated concentration of CaATP was kept constant at 1 μM. Under these conditions, Ca^{2+}-ATPase activity rises along the characteristic biphasic curve with respect to MgATP without any sign of inhibition by Ca^{2+}. Conversely, if the activity is measured at 1, 10, 100, and 500 μM CaATP at a fixed Ca^{2+} concentration of 5 μM and at concentrations of MgATP above those necessary to saturate the high-affinity site for ATP (see Section III, this chapter), CaATP inhibits the activity in a concentration-dependent fashion (Figure 9). Based on this, Muallem and Karlish[38] concluded that the inhibitory ligand is CaATP, acting at the low-affinity site where it displaced MgATP, which, in these authors' view, is the only effective ligand at that site. Although there are alternative explanations to account for the biphasic curve of Ca^{2+}-ATPase activity with respect to Ca^{2+}, the one advanced by Muallem and Karlish is attractive. Nevertheless, their proposal does not explain how Ca^{2+} inhibits almost 100% of the Ca^{2+}-ATPase activity, when if acting through the formation of CaATP complex, it should inhibit the fraction of activity due to occupation of the low-affinity site only, since the high-affinity site does not discriminate between free and complexed ATP. Regardless of the underlying mechanism, it should be remembered that the concentrations of Ca^{2+} necessary for the inhibition to become apparent are much higher than those in the cytosol, and hence it is unlikely that inhibition by intracellular Ca^{2+} would develop in a living cell under physiological conditions.

It is worth considering here some aspects of the Ca^{2+} curve of the Ca^{2+}-ATPase from sarcoplasmic reticulum, because it resembles that of the Ca^{2+} pump from plasma membrane. Ikemoto[44] measured the rate of hydrolysis of ATP by sarcoplasmic reticulum prepared from rabbit skeletal muscle and solubilized with Triton X-100® as a function of Ca^{2+} concentration, finding that it increases, reaches a plateau at 10^{-6} $M Ca^{2+}$, and then at 10^{-4} $M Ca^{2+}$ begins to diminish, being almost zero at 10^{-2} $M Ca^{2+}$. Looking for

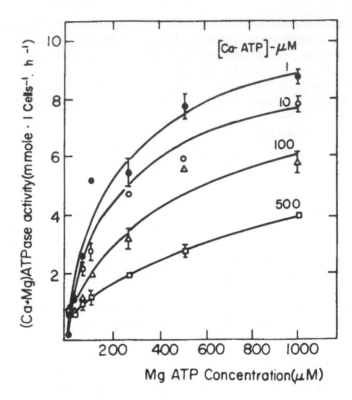

FIGURE 9.   Ca²⁺-ATPase activity from human red blood cell membranes as a function of MgATP at the concentrations of CaATP indicated on the curves and at constant (5 μM) Ca²⁺ concentration. (From Muallem, S. and Karlish, S. J. D., *Biochim. Biophys. Acta,* 647, 73, 1981. With permission.)

the mechanism responsible for such behavior, Ikemoto[44] found three types of noninteracting Ca²⁺-binding sites α, β, and γ with binding constants 4×10^{-6}, 4×10^{-4}, and 1×10^{-3} M, respectively. Binding of Ca²⁺ at the α site activates and binding at the γ site inhibits the Ca²⁺-ATPase. After these findings, the author concluded that there are sites intrinsic to the pump molecule from which Ca²⁺ regulates the rate of the Ca²⁺ pump from sarcoplasmic reticulum.

III. DEPENDENCE ON ATP

A. Substrate Specificity

For obvious reasons, the energy-donating substrate of the Ca²⁺ pump of plasma membrane is needed at the internal surface of the cell membrane. It is now accepted that the substrate is ATP, which is hydrolyzed to ADP and Pi, both of which are released into the cytosol. The specific requirements for ATP have been studied during Ca²⁺ transport as well as ATPase activity. In connection with the Ca²⁺-ATPase activity, we have found that in isolated human red blood cell membranes, ITP, CTP, GTP, or UTP do not substitute for ATP as the substrate.[45] Similar results have been reported for a purified Ca²⁺-ATPase from human erythrocytes.[46] Nucleotides other than ATP do not support active transport of Ca²⁺ in plasma membrane preparations of oviduct shell gland,[47] duodenum epithelia,[48] human lymphocyte,[49] and guinea pig leukocyte.[32] The ATP analog β-γ-(CH₂)-ATP does not serve as the substrate for Ca²⁺ transport.[50]

These series of experimental results give support to the idea accepted now that the

Ca²⁺ pump from plasma membranes uses specifically ATP as the energy-donating substrate. There are reports in the literature that do not agree with this view. The first studies on substrate specificity, published in 1969, stated that ATP, CTP, and UTP incorporated into resealed ghosts from human red blood cells are almost equally effective in supporting Ca²⁺ transport, GTP and ITP being somewhat less effective[51,52] These results, suggesting the lack of the specific requirement of ATP, have to be interpreted with care, since red blood cells contain a very active phosphoglycerate kinase which could have catalyzed phosphorylation of the ADP remaining in the ghosts to give enough ATP to energize the Ca²⁺ pump. Nevertheless, activation of Ca²⁺ transport by nucleoside triphosphates other than ATP has also been observed in IOVs under conditions in which synthesis of ATP should be negligible. In this respect, Sarkadi et al.[11] reported that with ITP or UTP as the substrate, IOVs from human red blood cells pump Ca²⁺ at a rate that is about half that attained with ATP, AMP being uneffective in this respect.

As mentioned previously, provided ATP is present, the Ca²⁺ pump from human red blood cell membranes hydrolyzes p-nitrophenylphosphate at a good rate. We have found that this hydrolysis is not coupled to the movement of Ca²⁺ across the membrane and hence that p-nitrophenylphosphate does not replace ATP as the energy source.[21] There is a marked difference between the Ca²⁺ pump from plasma membranes and that from sarcoplasmic reticulum in relation with the substrate specificity, because p-nitrophenylphosphate, acetylphosphate, as well as UTP, CTP, GTP, and ITP can substitute for ATP during Ca²⁺ transport in sarcoplasmic reticulum.[53] It seems that in terms of substrate requirements, the Ca²⁺ pump from plasma membrane is closer to the Na⁺ pump, because the latter needs ATP and cannot be energized by p-nitrophenylphosphate either.[54]

B. The Substrate Curve

In 1978, we tested human erythrocyte membranes for Ca²⁺-ATPase activity within a range of ATP concentration from 0.5 to 4000 μM.[55] We measured the release of Pi from (γ-³²P)ATP when the concentration of ATP was less than 100 μM and from nonradioactive ATP at higher nucleotide concentrations. Plotting the ATPase activity values against the concentration of ATP the biphasic curve shown in Figure 10 is obtained. Reciprocal plots of the Ca²⁺-ATPase activity against ATP concentration from 0.5 to 16 μM and from 0.125 to 4.0 mM yield straight lines as if they were two independent Michaelis-Menten curves (Figure 10). The response of the Ca²⁺-ATPase activity to ATP concentration therefore can be expressed by an equation of the form

$$v = v_1 + v_2 = \frac{Vm_1}{1 + \dfrac{Km_1}{[S]}} + \frac{Vm_2}{1 + \dfrac{Km_2}{[S]}} \tag{1}$$

where [S] is the concentration of ATP, v_1 represents the rate of the component observed at low ATP which shows high-apparent affinity for ATP (Km_1) and low-maximum velocity, and v_2 the rate of the component observed at high ATP with apparent affinity for ATP (Km_2) about 50 times lower and a maximum velocity 3 to 10 times higher than that of the first component. For the experiment in Figure 3 of Chapter 4, the kinetic parameters of Equation 1 are Km_1 2.5 μM, Km_2 143 μM, Vm_1 0.034 μmol/mg protein/hr, and Vm_2 0.375 μmol/mg protein/hr.

Muallem and Karlish[56] confirmed the biphasic response of the Ca²⁺-ATPase to ATP in resealed ghosts from human red blood cells and extended it to the active transport of Ca²⁺, demonstrating in the same preparation that the rates of ATP hydrolysis and transport of Ca²⁺ at increasing ATP concentrations follow biphasic kinetics. The rates

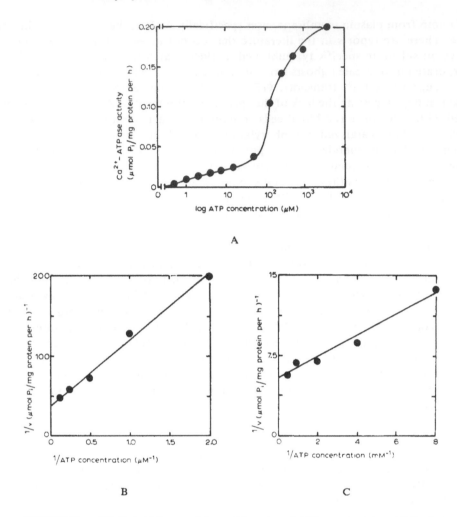

FIGURE 10.   (A) Ca^{2+}-ATPase activity as a function of ATP concentration. (B) Reciprocal plots of the values of Ca^{2+}-ATPase activity from against ATP concentrations in the 0.5-16 μM concentration range. (C) Reciprocal plots of the values of Ca^{2+}-ATPase activity from A against ATP concentrations in the 0.125-4 mM ATP concentration range. The value of the maximal velocity calculated from the plot in B was substrated from each of the values of Ca^{2+}-ATPase activity found in this ATP concentration range. (From Richards, D. E., Rega, A. F., and Garrahan, P. J., *Biochim. Biophys. Acta,* 511, 194, 1978. With permission.)

of hydrolysis and Ca^{2+} uptake by IOVs from human erythrocytes give also a biphasic response to ATP.[12,21] Biphasic ATP activation curves for the Ca^{2+} pump has also been found in membrane preparations from rat pancreatic islet cells[31] and human lymphocytes.[49] This finding is important because it is against the possibility of biphasic substrate curves being a characteristic of red cell membranes only.

Since all the results mentioned above were obtained with crude membrane preparations, it could be argued that biphasic curves could reflect the functioning of two ATP-fueled Ca^{2+} pumps of different Km for ATP. The main argument that can be raised against this interpretation comes from the finding by Stieger and Luterbacher that a highly purified preparation from human erythrocyte with more than 90% of its protein being made up by Ca^{2+} pump shows an ATP dependence with two Km values of 3.5 and 120 μM.[57] Hence, the complex kinetics during activation by ATP is indeed a property of the Ca^{2+} pump from plasma membranes.

Either calmodulin or Ca²⁺ have been shown to change the shape of the substrate curve of the Ca²⁺-ATPase.[58,59] Their effects are described in Chapter 10, Section I.B. The meaning of the complex response of the Ca²⁺-ATPase to ATP during ATP hydrolysis is analyzed in Chapter 8.

The existence of high-affinity together with low-affinity components during activation by ATP is well documented in the Na⁺ pump[60] as well as in the Ca²⁺ pump from sarcoplasmic reticulum,[61] so that biphasic response to the concentration of ATP seems a common property of the cation transport ATPases.

C. Kinetic Analysis of the Substrate Curve

It is tempting to consider that a sum of two Michaelis-Menten equations is the actual expression of the kinetic mechanism of the ATPase reaction and that the values of the two Vm's and the two Km's obtained from this equation have a definite physical meaning. We will show here that very different mechanisms predict biphasic response to the substrate and that the kinetic equations describing these mechanisms can be converted through a physically meaningless procedure into the sum of two Michaelis-Menten equations.

Therefore, equations derived from theoretical models that describe the actual mechanism will not give a better fit to experimental data than equations derived from models which are unrelated to the reaction mechanism of the Ca²⁺-ATPase.

1. Kinetic Schemes that Give Biphasic Substrate Curves

Biphasic substrate curves will appear in enzymes in which:

1. During each catalytic cycle, the substrate binds to the enzyme at two sites.
2. The sites have very different apparent affinities. Obviously one of them must be the catalytic site. The other may be either a catalytic or a regulatory site at which the substrate acts as an activator without being transformed.
3. If the enzyme follows rapid-equilibrium kinetics, the two sites for substrate have to be physically distinct. Under steady-state conditions, the sites can be different states of a single site during a catalytic cycle.
4. To obtain a biphasic response, it is mandatory that activity be also possible when only one site is occupied by the substrate. If this were not so, the relation between activity and substrate concentration would be sigmoid instead of biphasic.
5. The activity observable when only one site is occupied must necessarily include that corresponding to the site of higher affinity. Hence, when one of the sites is not catalytic, the catalytic site must be the site of high affinity and when the differences in affinity result from interactions among otherwise identical sites the interactions must be negative, i.e., the affinity of a site must be higher when the other site is unoccupied.

In what follows, we will briefly describe the main particular cases in which the conditions outlined above are fulfilled.

a. Two Different Enzymes

The simplest case is that of enzyme preparations containing two noninterconverting enzymes differing significantly in their Km values. In this case, both steady-state and rapid-equilibrium kinetics predict that the substrate curve will be the sum of the contribution of the Michaelis-Menten equations that define the kinetics of each enzyme such as that shown in Equation 1.

b. Two Active Sites in the Same Enzyme

Both under steady-state or rapid-equilibrium conditions, an equation identical to

Equation 1 would describe the behavior of an enzyme if two different populations of noninteracting active sites for the substrate were present in each enzyme.

If the affinity of a site depended on the occupation of the rest, the kinetic equation would no longer be a sum of two Michaelis-Menten equations. For example, if the sites had the same intrinsic affinity in the absence of substrate and rapid-equilibrium held for the addition of the substrate, the kinetic equation would have the form:

$$v = \frac{Vm_1}{1 + \dfrac{Ks}{[S]} + \dfrac{Ks[S]}{2\,Ks'}} + \frac{Vm_2}{1 + \dfrac{2Ks'}{Ks[S]}\left(1 + \dfrac{Ks}{[S]}\right)} \tag{2}$$

where Ks and Ks' are the equilibrium constants for the dissociation of S from one site when the other is empty or occupied, respectively, and Vm_1 and Vm_2 are the maximum rates attainable when only one or both sites, respectively, are saturated. Equation 2 predicts a biphasic response when $Ks \ll Ks'$.

c. The Substrate as Activator

If an enzyme had only one class of active site, and catalysis proceeded slowly when the substrate was bound to this site and quickly when the substrate was also bound at a noncatalytic site, the following equation would describe the response to the substrate under rapid-equilibrium conditions

$$v = \frac{Vm_1}{1 + \dfrac{Ks_1\,Ks_2}{(Ks_1 + Ks_2)\,[S]} + \dfrac{Ks_2\,[S]}{Ks_2'\,(Ks_2 + Ks_1)}}$$
$$+ \frac{Vm_2}{1 + \dfrac{(Ks_1' + Ks_2')}{[S]}\left(1 + \dfrac{Ks_1\,Ks_2}{(Ks_1' + Ks_2')\,[S]}\right)} \tag{3}$$

where Ks_1 and Ks_1' are the dissociation constants of the substrate from the catalytic site when the site for the activator is free or occupied, respectively, Ks_2 and Ks_2' are the dissociation constants of the substrate from the activator site when the catalytic site is free or occupied, respectively, Vm_1 is the maximum velocity when only the catalytic site is occupied, and Vm_2 the maximum velocity when both sites are occupied.

If there were no interactions between the affinities of the sites $Ks_1 = Ks_1'$ and $Ks_2 = Ks_2'$, Equation 3 would become:

$$v = \frac{Vm_1}{1 + \dfrac{Ks_1}{[S]}} + \frac{(Vm_2 - Vm_1)}{\left(1 + \dfrac{Ks_1}{[S]}\right)\left(1 + \dfrac{Ks_2}{[S]}\right)} \tag{4}$$

We have already mentioned that under the rapid equilibrium assumption, biphasic substrate curves are impossible in an homogenous population of enzymes having a single class of sites. This is so, because the rapid equilibrium assumption implies that the velocity vs. substrate concentration curve will differ only in a constant factor from the equilibrium binding vs. substrate concentration curve and biphasic binding curves are only seen when more than one class of site existing simultaneously are involved in the binding process.

Under steady-state kinetics, changes in the properties of a single site during a catalytic cycle can generate biphasic substrate curves.

To illustrate this, we will take as example the substrate kinetics of the (Na^+, K^+)-ATPase. In this system, there is direct experimental evidence that apart from acting as the substrate, ATP accelerates with low affinity and without being hydrolyzed.[60] This effect is exerted on the site that in other conformers of the enzyme acts as the high-affinity catalytic site. Under these conditions, the steady-state rate equation for initial velocity in the absence of products and at nonlimiting concentrations of all ligands but ATP would be:

$$v = \frac{Vm_1}{1 + \dfrac{Km_1}{[S]} + \dfrac{[S]}{Km_2}} + \frac{Vm_2}{1 + \dfrac{Km_2}{[S]}\left(1 + \dfrac{Km_1}{[S]}\right)} \tag{5}$$

2. Comparison of the Kinetic Equations

From the preceding analysis, it seems clear that rate equations that are the sum of two Michaelis-Menten expressions will have physical meaning only for those kinetic schemes that involve simultaneous catalysis at two different and independent active sites.

However, if the general conditions stated at the beginning of this section are fulfilled, for all the kinetic schemes we have analyzed, substrate curves will be biphasic with low- and high-apparent affinity components. Depending on the particular kinetic scheme, the differences in affinity will be either intrinsic (Equation 1 and 4) or result from anticooperative interactions between different sites (Equations 2 and 3), or from changes in the affinity of a single site during the catalytic cycle (Equation 5).

a. The Mathematical Equivalence of Rate Equations

If there are large differences between the apparent affinities of the sites, at sufficiently low concentrations of substrate, in all the rate equations the terms containing the ratio:

[S]/low-affinity apparent dissociation constant

will be near zero and the terms containing the ratio:

Low-affinity apparent dissociation constant/[S]

will be much smaller than 1.

Under these conditions, as the concentration of substrate decreases, the rate equations for all the kinetic schemes we have analyzed will tend to the sum of a Michaelis-Menten equation plus a straight line which will have the form:

$$v = \frac{Vm_1}{1 + \dfrac{Km_1}{[S]}} + \frac{Vm_2}{Km_2}[S] \tag{6}$$

for Equation 1, and the form:

$$v = \frac{[Vm_1 - Vm_2\,(Km_1/Km_2)]}{1 + \dfrac{Km_1}{[S]}} + \frac{Vm_2}{Km_2}[S] \tag{7}$$

for Equations 2 to 5.

Km_1 and Km_2 in Equations 6 and 7 describe the high- and low-affinity components of the rate equations. The physical meaning of these parameters will differ, depending on the kinetic mechanism.

When the concentration of the substrate is sufficiently high, the ratios:

High-affinity apparent dissociation constant/[S]

will approach zero and all kinetic equations will tend to the sum of a constant term plus a Michaelis-Menten equation with the form:

$$v = Vm_1 + \frac{Vm_2}{1 + \frac{Km_2}{[S]}} \qquad (8)$$

for Equation 1, and the form:

$$v = Vm_1 + \frac{Vm_2 - Vm_1}{1 + \frac{Km_2}{[S]}} \qquad (9)$$

for Equations 2 to 5 in which Km_2 has the meaning explained when describing Equations 6 and 7.

Notice that the rate at nonlimiting substrate concentrations is the sum of Vm_1 and Vm_2 for those models whose mechanistic equations are the sum of two Michaelis-Menten-like terms, whereas it is equal to Vm_2 for those which are not mechanistically described by the sum of two Michaelis-Menten-like terms.

Equations 6 to 9 demonstrate that at sufficiently low or high concentration of substrate the rate equations describing the different kinetic schemes we have analyzed become mathematically indistinguishable.

This is a particular example of a much general mathematical equivalence which can be stated as follows: all kinetic equations we have described can be written as the quotient of two second-degree polynomials, i.e.:

$$v = \frac{A[S]^2 + B[S]}{C[S]^2 + D[S] + E} \qquad (10)$$

in which the physical meaning of the coefficients A, B, C, D, and E will depend on the underlying kinetic mechanism.

If all equations can be expressed by Equation 10, then all must be interconvertible. For this reason, experimental data from systems governed by one of the kinetic equations we have described will fit to any other of the kinetic equations, including the sum of two Michaelis-Menten equations, with exactly the same degree of precision as the equation that described the actual kinetics of the system. The mathematical equivalence of the equations does not imply that the physical models are equivalent. For this reason, only when the particular solution of a kinetic model yields a sum of two Michaelis-Menten equations will the parameters of these equations have a definite physical meaning.

We can conclude, therefore, that in most of the cases, curve fitting cannot be used as an experimental tool to decide which of the different kinetic mechanisms that generate biphasic substrate curves is reponsible for this phenomenon in the Ca^{2+}-ATPase. Other criteria have to be applied to solve this problem.

D. On the State of ATP as the Substrate for the Overall Reaction

The subject has been a matter of dispute and an unquestionable agreement among the authors has not been reached yet. Moreover, the existence of two kinetically distinct sites for ATP represents an additional complication for the matter, since ATP species that are effective at the high-affinity site could be ineffective at the low-affinity site.

Since the Ca^{2+} pump requires only ATP, Mg^{2+}, and Ca^{2+} to express all its activities, the possibilities are that the state of ATP as the substrate of the pump is either (1) free-ATP, (2) CaATP, (3) MgATP, or (4) any of the three species. Most of the information on the subject comes from experiments made on human red blood cell membranes. We have proposed that free-ATP can serve as the substrate of the Ca^{2+} pump after finding that, although at a low rate, Ca^{2+} ATPase activity from human red blood cell membranes persists in the absence of added Mg^{2+}.[55,62] One important complementary factor favoring free-ATP as the substrate comes from measurements of Ca^{2+}-ATPase activity made by Schatzmann.[63] He found that the Km is independent of Ca^{2+}, and assuming that free-ATP rather than MgATP is the substrate, the Km value during the overall ATPase reaction is 1 to 2 μM, just about the value obtained by us for the Km of the phosphorylation reaction.[64] Experimental evidence that CaATP may serve as the substrate during the Ca^{2+}-ATPase reaction comes mainly from Penniston's laboratory. They measured Ca^{2+} uptake by IOVs from red cells as a function of increasing calcium concentration at 1 μM ATP without added magnesium. Under these circumstances, Ca^{2+} uptake increases, while free-ATP is driven to extremely low levels.[65] The Km for CaATP is 0.010 μM, a low value indeed, but coincidentally it is just about the concentration of CaATP that can be calculated to be present in the cytosol of a red blood cell at pH 7.2.[66] Although it is not proof that CaATP may serve as the substrate for the Ca^{2+}-ATPase reaction, it seems worth mentioning here that maximum steady-state levels of phosphorylation are attained in media containing 2 μM total ATP, 10 μM magnesium, and 10 mM calcium, conditions under which practically all the ATP will be CaATP.[67] Measuring Ca^{2+}-ATPase activity at different pH values, Wolf[68] found that the data can be fitted assuming that MgATP, but not free-ATP, is the substrate. Using IOVs, Sarkadi et al.[69,70] found that at ATP and total calcium concentrations near 20 μM active Ca^{2+} uptake increases with magnesium in parallel with the calculated concentration of MgATP, and the rate is maximum at 1000 μM added magnesium, conditions in which the concentration of either free-ATP or CaATP can be calculated to be less than 2 μM. Muallem and Karlish[71] measured Ca^{2+}-ATPase activity of human red blood cell membranes and found that within the range 100 to 2000 μM, MgATP accelerates ATP hydrolysis by combining at the low-affinity site for ATP, and CaATP behaves as a mostly competitive inhibitor at this site. Free-ATP does not appear to interfere with MgATP. Muallem and Karlish[71] also found that at 5 μM MgATP, conditions under which the activity is mostly due to combination of the nucleotide at the high-affinity site only, the Ca^{2+}-ATPase activity is independent of both CaATP within 1 to 500 μM and free-ATP within 13 to 6000 μM, so that any of the three mentioned ATP species may serve as the substrate of the Ca^{2+} pump.

It is difficult to draw a definite conclusion on the nature of the substrate for the Ca^{2+} pump from plasma membranes. However, from the data available now, it seems that as proposed by Muallem and Karlish[56] the Ca^{2+} pump does not discriminate between MgATP, CaATP, and free-ATP at the high-affinity site, and hence that any of the three species serves as the substrate of the system. It has also to be mentioned that the requirements for ATP at the high- and the low-affinity sites may be different and that MgATP appears to be the most likely candidate for the low-affinity site.[66] This is treated in more detail in connection with the effects of Mg^{2+} in Chapter 9.

REFERENCES

1. Schatzmann, H. J., Dependence on calcium concentration and stoichiometry of the calcium pump in human red cells, *J. Physiol.*, 235, 551, 1973.
2. Garrahan, P. J. and Rega, A. F., Cation loading of red blood cells, *J. Physiol.*, 193, 459, 1967.
3. Ferreira, H. G. and Lew, V. L., Use of ionophore A23187 to measure cytoplasmic Ca buffering and activation of the Ca pump by internal Ca, *Nature*, 259, 47, 1976.
4. Sarkadi, B., Szász, I., Gerlóczy, A., and Gárdos, G., Transport parameters and stoichiometry of active calcium extrusion in intact human red cells, *Biochim. Biophys. Acta*, 464, 93, 1977.
5. Simonsen, L. O., Gomme, J., and Lew, V. L., Uniform ionophore A23187 distribution and cytoplasmic calcium buffering in intact human red cells, *Biochim. Biophys. Acta*, 692, 431, 1982.
6. Tsien, R. Y., A non-disruptive technique for loading calcium buffers and indicators into cells, *Nature*, 290, 527, 1981.
7. Kratje, R. B., Garrahan, P. J., and Rega, A. F., The effects of alkali metal ions on active Ca²⁺ transport in reconstituted ghosts from human red cells, *Biochim. Biophys. Acta*, 731, 40, 1983.
8. Muallem, S. and Karlish, S. J. D., Is the red cell calcium pump regulated by ATP?, *Nature*, 277, 238, 1979.
9. Steck, T. L. and Kant, J. A., Preparation of impermeable ghosts and inside-out vesicles from human erythrocyte membranes, *Methods in Enzymology*, 31, 172, 1974.
10. Lew, V. L., Muallem, S., and Seymour, C. A., One-step vesicles from mammalian red cells, *J. Physiol.*, 307, 36P, 1980.
11. Sarkadi, B., Szász, I., and Gárdos, G., Characteristics and regulation of active calcium transport in inside-out red cell membrane vesicles, *Biochim. Biophys. Acta*, 598, 326, 1980.
12. Mollman, J. E. and Pleasure, D. E., Calcium transporting human inside-out erythrocyte vesicles, *J. Biol. Chem.*, 255, 569, 1980.
13. Haaker, H. and Racker, E., Purification and reconstitution of the Ca²⁺-ATPase from plasma membranes of pig erythrocytes. *J. Biol. Chem.*, 254, 6598, 1979.
14. Carafoli, E. and Zurini, M., The Ca²⁺-pumping ATPase of plasma membranes. Purification, reconstitution and properties, *Biochim. Biophys. Acta*, 683, 279, 1982.
15. Caroni, P., Zurini, H., Clark, A., and Carafoli, E., Further characterization and reconstitution of the purified Ca²⁺ pumping ATPase of heart sarcolemma, *J. Biol. Chem.*, 258, 7305, 1983.
16. DiPolo, R. and Beaugé, L., Mechanisms of calcium transport in the giant axon of the squid and their physiological role, *Cell Calcium*, 1, 147, 1980.
17. Schatzmann, H. J., Dependence on calcium concentration and stoichiometry of the calcium pump in human red cells, *J. Physiol.*, 235, 551, 1973.
18. Sarkadi, B., Enyedi, A., Szász, I., and Gárdos, G., Effects of calmodulin on active calcium uptake and membrane phosphorylation in inside-out red cell membrane vesicles, *Adv. Physiol. Sci.*, 6, 181, 1981.
19. Waisman, D. M., Gimble, J. M., Goodman, D. B. P., and Rasmussen, H., Studies of the Ca²⁺ transport mechanism of human erythrocyte inside out plasma membrane vesicles. I. Regulation of the Ca²⁺ pump by calmodulin, *J. Biol. Chem.*, 256, 409, 1981.
20. Cha, Y. N., Shin, B. C., and Lee, K. S., Active uptake of Ca²⁺ and Ca²⁺-activated Mg²⁺ ATPase in red cell membrane fragments, *J. Gen. Physiol.*, 202, 47, 1971.
21. Caride, A. J., Rega, A. F., and Garrahan, P. J., Effects of p-nitrophenylphosphatase on Ca²⁺ transport in inside out vesicles from human red cell membranes, *Biochim. Biophys. Acta*, 734, 363, 1983.
22. Gimble, J. M., Waisman, D. M., Gustin, J., Goodman, D. B. P., and Rasmussen, H., Studies of the Ca²⁺ transport mechanism of human erythrocyte inside out membrane vesicles. Evidence for the development of a positive interior membrane potential, *J. Biol. Chem.*, 257, 10781, 1982.
23. Waisman, D. M., Gimble, J. M., Goodman, D. B. P., and Rasmussen, H., Studies of the Ca²⁺ transport mechanism of human erythrocyte inside out plasma membrane vesicles. II. Stimulation of the Ca²⁺ pump by phosphate, *J. Biol. Chem.*, 256, 415, 1981.
24. Waisman, D. M., Gimble, J. M., Goodman, D. B. P., and Rasmussen, H., Studies of the Ca²⁺ transport mechanism of human erythrocyte inside out plasma membrane vesicles. III. Stimulation of the Ca²⁺ pump by anions, *J. Biol. Chem.*, 256, 420, 1981.
25. Waisman, D. M., Smallwood, J., Lafrenier, D., and Rasmussen, H., The role of band III in calcium transport across the human erythrocyte membrane, *FEBS Lett.*, 145, 337, 1982.
26. Rossi, J. P. F. C. and Schatzmann, H. M., Is the red cell calcium pump electrogenic?, *J. Physiol.*, 327, 1, 1982.
27. Zimmiak, P. and Racker, E., Electrogenicity of Ca²⁺ transport catalyzed by the Ca²⁺-ATPase from sarcoplasmic reticulum. *J. Biol. Chem.*, 253, 4631, 1978.
28. Niggli, V., Sigel, E., and Carafoli, E., The purified Ca²⁺ pump of human erythrocyte membranes catalyzes an electroneutral Ca²⁺- H⁺ exchange in reconstituted liposomal systems, *J. Biol. Chem.*, 257, 2350, 1982.

29. Smallwood, J. I., Waisman, D. M., Lafrenier, D., and Rasmussen, H., Evidence that the erythrocyte calcium pump catalyzes a $Ca^{2+}:nH^+$ exchange, *J. Biol. Chem.*, 258, 11092, 1983.

30. Schatzmann, H. J. and Roelofsen, B., Some aspects of the Ca pump in human red blood cells, in *Biochemistry of Membrane Transport,* Semenza, G. and Carafoli, E., Eds., Springer-Verlag, Berlin, 389, 1977.

31. Pershadsingh, H. A., McDaniel, M. L., Landt, M., Bry, C. G., Lacy, P. E., and McDonald, J. M., Ca^{2+}-activated ATPase and ATP-dependent calmodulin-stimulated Ca^{2+} transport in islet cell plasma membrane, *Nature,* 288, 492, 1980.

32. Ochs, D. L. and Reed, P. W., ATP-dependent calcium transport in plasma membrane vesicles from neutrophil leukocytes, *J. Biol. Chem.,* 258, 10116, 1983.

33. Breitbart, H., Stern, B., and Rubinstein, S., Calcium transport and Ca^{2+}-ATPase activity in ram spermatozoa plasma membrane vesicles, *Biochim. Biophys. Acta,* 728, 349, 1983.

34. Klaven, N. B., Pershadsingh, H. A., Henius, G. V., Laris, P. C., Long, J. W., and McDonald, J. M., A high-affinity, calmodulin-sensitive (Ca^{2+} + Mg^{2+})-ATPase and associated calcium-transport pump in the Ehrlich ascites tumor cell plasma membrane, *Arch. Biochem. Biophys.,* 226, 618, 1983.

35. Scharf, O., Regulation of (Ca^{2+}, Mg^{2+})-ATPase in human erythrocytes dependent on calcium and calmodulin, *Acta Biol. Med. Ger.,* 40, 457, 1981.

36. Downes, P. and Michell, R. H., Human erythrocyte membranes exhibit a cooperative calmodulin-dependent Ca^{2+}-ATPase of high calcium sensitivity, *Nature,* 290, 270, 1981.

37. Wutrich, A., unpublished, Cited by Schatzmann, H. J., The plasma membrane calcium pump of erythrocytes and other animal cells, in *Membrane Transport of Calcium,* Carafoli, E., Ed., Academic Press, New York, 1982, chap. 2.

38. Muallem, S. and Karlish, S. J. D., Studies on the mechanism of regulation of the red cell Ca^{2+} pump by calmodulin and ATP, *Biochim. Biophys. Acta,* 647, 73, 1981.

39. Roufogalis, B. D., Akyempon, Ch. K., Al-Jobore, A., and Minocherhomjee, A. M., Regulation of the Ca^{2+} pump of the erythrocyte membrane, *Ann. N.Y. Acad. Sci.,* 402, 349, 1982.

40. Gmaj, P., Murer, H. and Carafoli, E., Localization and properties of a high-affinity (Ca^{2+} + Mg^{2+})-ATPase in isolated kidney cortex plasma membranes, *FEBS Lett.,* 144, 226, 1982.

41. Niggli, V., Adunyah, E. S., Cameron, B. F., Bababunni, E. Z., and Carafoli, E., The Ca^{2+}-pump of sickle cell plasma membranes. Purification and reconstitution of the ATPase enzyme, *Cell Calcium,* 3, 131, 1982.

42. Schatzmann, H. J., The plasma membrane calcium pump of erythrocytes and other animal cells, in *Membrane Transport of Calcium,* Carafoli, E., Ed., Academic Press, New York, 1982.

43. Lichner, R. and Wolf, H. U., Phosphorylation of the isolated high-affinity (Ca^{2+} + Mg^{2+})-ATPase of the human erythrocyte membrane, *Biochim. Biophys. Acta,* 598, 472, 1980.

44. Ikemoto, N., The calcium binding sites involved in the regulation of the purified adenosine triphosphatase of the sarcoplasmic reticulum. *J. Biol. Chem.,* 249, 649, 1974.

45. Rega, A. F., Richards, D. E., and Garrahan, P. J., Calcium ion-dependent *p*-nitrophenylphosphate phosphatase activity and calcium ion-dependent adenosine triphosphatase activity from human erythrocyte membranes, *Biochem. J.,* 136, 185, 1973.

46. Graf, E., Verna, A. K., Gorski, J. P., Lopaschuk, G., Niggli, V., Zurini, M., Carafoli, E., and Penniston, J. T., Molecular properties of calcium-pumping ATPase from human erythrocytes, *Biochemistry,* 21, 4511, 1982.

47. Coty, W. A. and McConkey, C., Jr., A high-affinity calcium-stimulated ATPase activity in the hen oviduct shell gland, *Arch. Biochem. Biophys.,* 219, 444, 1982.

48. Ghijsen, W. E. J. M. and Van Os, C. H., Ca-stimulated ATPase in brush border and basolateral membranes of rat duodenum with high affinity sites for Ca ions, *Nature,* 279, 802, 1979.

49. Lichtman, A. H., Segel, G. B., and Lichtman, M. A., Calcium transport and calcium-ATPase activity in human lymphocyte plasma membrane vesicles, *J. Biol. Chem.,* 256, 12, 1981.

50. Ghijsen, W. E. J. M., De Jong, M. D., and Van Os, C. H., ATP-dependent calcium transport and its correlation with Ca^{2+}-ATPase activity in basolateral plasma membranes of rat duodenum, *Biochim. Biophys. Acta,* 689, 327, 1982.

51. Olson, E. J. and Cazort, R. J., Active calcium and strontium transport in human erythrocyte ghosts, *J. Gen. Physiol.,* 53, 311, 1969.

52. Lee, K. S. and Skin, B. C., Studies on the active transport of calcium in human red cells, *J. Gen. Physiol.,* 54, 713, 1969.

53. de Meis, L., The sarcoplasmic reticulum. Transport and energy transduction, in *Transport in the Life Sciences,* Vol. 2, Bittar, E. E., Ed., John Wiley & Sons, New York, 1980.

54. Garrahan, P. J. and Rega, A. F., Potassium activated phosphatase from human red blood cells. The effects of *p*-nitrophenylphosphate on cation fluxes, *J. Physiol.,* 223, 595, 1972.

55. Richards, D. E., Rega, A. F., and Garrahan, P. J., Two classes of site for ATP in the Ca^{2+}-ATPase from human red cell membranes, *Biochim. Biophys. Acta,* 511, 194, 1978.

56. Muallem, S. and Karlish, S. D., Is the red calcium pump regulated by ATP?, *Nature,* 277, 238, 1979.

57. Stieger, J. and Luterbacher, S., Some properties of the purified (Ca^{2+} + Mg^{2+})-ATPase from human red cell membranes, *Biochim. Biophys. Acta,* 641, 270, 1981.

58. Muallem, S. and Karlish, S. J. D., Regulatory interaction between calmodulin and ATP on the red cell Ca^{2+} pump, *Biochim. Biophys. Acta,* 597, 631, 1980.

59. Scharf, O., Kinetics of calmodulin-dependent (Ca^{2+} + Mg^{2+})-ATPase in plasma membranes and solubilized membranes from erythrocytes, *Arch. Biochem. Biophys.,* 209, 72, 1981.

60. Robinson, J. D. and Flashner, M. S., The (Na$^+$ + K$^+$)-activated ATPase. Enzymatic and transport properties, *Biochim. Biophys. Acta,* 549, 145, 1979.

61. Tada, M., Yamamoto, T., and Tonomura, Y., Molecular mechanism of active calcium transport by sarcoplasmic reticulum, *Physiol. Revs.,* 58, 1, 1978.

62. Garrahan, P. J. and Rega, A. F., Activation of the partial reactions of the Ca^{2+}-ATPase of human red cells by Mg^{2+} and ATP, *Biochim. Biophys. Acta,* 513, 59, 1978.

63. Schatzmann, H. J., Role of magnesium in the (Ca^{2+} + Mg^{2+})-stimulated membrane ATPase of human red blood cells, *J. Membrane Biol.,* 35, 149, 1977.

64. Rega, A. F. and Garrahan, P. J., Calcium dependent phosphorylation of human erythrocyte membranes, *J. Membrane Biol.,* 22, 313, 1975.

65. Penniston, J. T., Substrate specificity of the erythrocyte Ca^{2+}-ATPase, *Biochim. Biophys. Acta,* 688, 735, 1982.

66. Graf, E. and Penniston, J. T., CaATP: the substrate, at low ATP concentrations, of Ca^{2+}-ATPase from human erythrocyte membranes, *J. Biol. Chem.,* 256, 1587, 1981.

67. Lichtner, R. and Wolf, H. V., Phosphorylation of the isolated high-affinity (Ca^{2+} + Mg^{2+})-ATPase of the human erythrocyte membrane, *Biochim. Biophys. Acta,* 598, 472, 1980.

68. Wolf, H. U., Studies on a Ca^{2+}-dependent ATPase of human erythrocyte membranes. Effects of Ca^{2+} and H$^+$, *Biochim. Biophys. Acta,* 266, 361, 1972.

69. Sarkadi, B., Enyedi, A., and Gárdos, G., Metal-ATP complexes as substrates and free metal ions as activators of the red cell calcium pump, *Cell Calcium,* 2, 449, 1981.

70. Enyedi, A., Sarkadi, B., and Gárdos, G., On the substrate specificity of the red cell calcium pump, *Biochim. Biophys. Acta,* 687, 109, 1982.

71. Muallem, S. and Karlish, S. J. D., Studies on the mechanism of regulation of the red-cell Ca^{2+} pump by calmodulin and ATP, *Biochim. Biophys. Acta,* 647, 73, 1981.

Chapter 7

OTHER PROPERTIES AND COUPLING OF Ca²⁺ TRANSPORT AND ATP HYDROLYSIS

A. F. Rega

I. OTHER PROPERTIES

A. Specificity for Ca²⁺

The selectivity of the Ca²⁺ pump from plasma membranes for Ca²⁺ is not so high, since other divalent metal ions can replace Ca²⁺ in the transport cycle. Furthermore, unpublished experimental results from our laboratory suggest that the site for Ca²⁺ in the transport ATPase can be occupied by Mg²⁺. Under these conditions, the activity of the enzyme ceases, because it does not accept Mg²⁺ for transport.

In 1969, Schatzmann and Vincenzi[1] and Alson and Cazort[2] found that in resealed ghosts, Sr²⁺ is transported out in the same way as Ca²⁺. The first authors also showed that the transport of Sr²⁺ can be abolished, increasing the concentration of Ca²⁺ inside the cell as if Ca²⁺ displaced Sr²⁺ from its site in the Ca²⁺ pump. They observed that in the presence of magnesium, although with less affinity, Sr²⁺ also replaces Ca²⁺ in activating the ATPase. This series of findings have been extended to IOVs from human red blood cells which in media with ATP plus Mg²⁺ take up Sr²⁺,[3] a process that is stimulated by calmodulin. Active Sr²⁺ uptake is competitively inhibited by Ca²⁺.[3]

The effects of divalent cations other than Ca²⁺ have also been studied on the Ca²⁺-ATPase activity. Hydrolysis of ATP by a purified preparation of the enzyme from human red blood cell membranes is stimulated by alkaline metal earth cations in the order Ca²⁺ > Sr²⁺ ≫ Ba²⁺.[4] Sr²⁺ is almost as effective as Ca²⁺ but its apparent affinity is lower. Pfleger and Wolf[5] studied with much detail the overall ATPase activity of red blood cell membranes (not necessarily belonging to the Ca²⁺ pump) in the presence of a series of divalent metal ions. They found that apart from Ca²⁺ and Sr²⁺, although to a lesser extent, Pb > Ni, Cu, Zn, Mn, Cd > Co > Ba in the presence of Mg²⁺ stimulate the ATPase (but see below). The plot of the ratio between the maximum activation for each of the cations tested and the maximum activation by Ca²⁺ vs. the ion radius is a bell-shaped curve (Figure 1) for all the cations except Co²⁺ and Hg²⁺, which fall down the curve, possibly because of their reaction with essential −SH groups which causes inactivation of the ATPase. Pfleger and Wolf also showed that there is no correlation between the activation potency of the divalent cations and the dissociation constants of the divalent metal ions-enzyme complex. After these findings, they concluded that the degree of effectiveness of divalent metal ions as activators of the ATPase depends on the ion radius rather than on the nature of the metal.

Although Mn²⁺, in the presence of Mg²⁺, activates the ATPase activity of human red blood cells, it is not transported and does not inhibit Ca²⁺ transport when both Ca²⁺ and Mg²⁺ are present inside resealed ghosts at the same concentration.[6] Hence, results on activation of ATPase activity by divalent metal ions other than Ca²⁺ or Sr²⁺ must be interpreted with care until demonstration that they correlate well with the active transport of the cation.

B. The Apparent Affinity for Ca²⁺

Since Ca²⁺ is transported towards the outside of the cell, for obvious reasons the site at which it combines to the Ca²⁺ pump is on the internal surface of the plasma membrane.

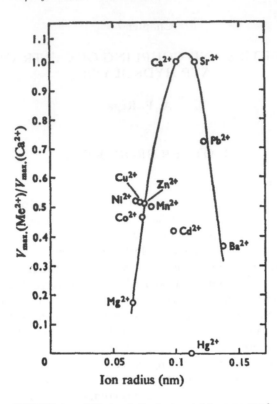

FIGURE 1.   Effects of divalent metal ions on total
ATPase activity from human red cell membranes in the
presence of 2 mM Mg^{2+} as a function of the ion radius.
On the ordinate, the ratio between the maximum activation for each of the cation and the maximum activation by Ca^{2+} is represented. (From Pfleger, H. and
Wolf, H. U., *Biochem. J.*, 147, 359, 1975. With permission.)

Up to now, no direct measurement of binding of Ca^{2+} to the site from which it is
transported by the Ca^{2+} pump has been reported so that all the information on it comes
from kinetic measurements. The apparent affinity (apparent dissociation constant) for
Ca^{2+} is taken as the concentration of Ca^{2+} necessary for half-maximal stimulation (K_{Ca})
of one of the activities of the pump. The energy of binding seems to be relatively high
since K_{Ca} is always in the micromolar range. For micromolar concentrations of Ca^{2+} in
solutions of complex composition to be kept constant during the duration of an experiment, they have to be buffered with chelators of Ca^{2+}. Often the final concentration
of Ca^{2+} in the buffer solution is not measured, but rather assumed to be that calculated
from the total concentration of calcium, magnesium, ATP, ADP, Pi, the chelator of
Ca^{2+}, etc., and the corresponding association constants taking account of the pH. Furthermore, when resealed ghosts are used, the internal concentration of Ca^{2+} in the
ghosts is assumed to be identical to that in the medium in which the ghosts were sealed,
regardless of the changes in cell volume that could occur after sealing and the binding
of Ca^{2+} to the cell components remaining within the ghosts.

In spite of the factors mentioned above that could give rise to uncertainties, the fact
is that all values that have been reported for the K_{Ca} of the Ca^{2+} pump during activation
of ATP hydrolysis and Ca^{2+} transport are in the micromolar range. Table 1 of Chapter
4 shows values of $K_{0.5}$ during Ca^{2+}-ATPase activity that have been found in plasma
membranes of different tissues but the red blood cell. The mean value is 0.696 μM with
a minimum of 0.09 μM for rat pancreas and a maximum of 4.6 μM for rat spermato-

zoa, being the latter the only one above $10^{-6} M$. Concerning the red blood cell, the K_{Ca} has been measured during Ca^{2+} transport in different preparations of human red blood cells and the values are (μM): fresh intact red blood cells 1;[7] resealed ghosts loaded with buffered calcium 4,[8] 3.3,[9] and IOVs 3.4.[10] Table 1 of Chapter 4 shows that in purified preparations of Ca^{2+} pump from human red blood cells, where it cannot be argued that the accessibility to the Ca^{2+} site is hampered by the other membrane components, the K_{Ca} is 1.5 and 20 μM in the presence and in the absence of calmodulin, respectively.

The fact is that these values are higher than most of those shown in Table 1 of Chapter 4 for tissues other than the red blood cell. Whether the observed difference reflects either differences among procedures used to estimate free calcium concentrations or a real lower affinity of the Ca^{2+} pump in red blood cells than in other tissues, is not known. Furthermore, is this apparent affinity of the Ca^{2+} pump from red cells compatible with a system that sustains intracellular concentrations of Ca^{2+} below $20^{-6} M$? Assuming a maximum rate of Ca^{2+} transport in intact cells of 20 mmol/ℓ cell/hr (disregarding the activation by K^+), K_{Ca} 5 μM and simple hyperbolic kinetics during activation by Ca^{2+}, the rate of Ca^{2+} efflux at 0.1 $\mu M\,Ca^{2+}$ in the cytosol (a concentration which does not hurt the cell) can be calculated to reach about 0.4 mmol/ℓ cells/hr. This simple calculation allows one to conclude that, although low, the rate of efflux under these conditions is 10 times higher than that necessary to compensate for the Ca^{2+} that enters into a red cell (see Chapter 2) which at most is 0.03 mmol/ℓ cells/hr.

1. Modifiers of the Apparent Affinity for Ca^{2+}

The high apparent affinity of the transport site for Ca^{2+} is independent of ligands like ATP and monovalent cations that participate in the transport cycle, but it can be modified by several substances and treatments. Among them, the acidic lipids, partial proteolysis, and calmodulin increase the apparent affinity. Their effects and possible mechanism of action are described in Chapter 10.

Mg^{2+}, an essential cofactor for the transport reaction, lowers the apparent affinity for Ca^{2+} during Ca^{2+}-ATPase activity in IOVs from red blood cells.[11] Unpublished experimental results from our laboratory show that Mg^{2+} increases K_{Ca} by displacing Ca^{2+} from its site in the Ca^{2+}-ATPase. In view of this, it is clear that to know the true apparent affinity, a series of K_{Ca} values at different Mg^{2+} concentrations should be obtained, the true K_{Ca} being that calculated by extrapolation at zero Mg^{2+} concentration.

Certain anionic and polyanionic compounds like salicylic and benzoic acid and poly(L-aspartic acid) or poly(L-glutamic acid) at concentrations near 100 μM increase the apparent affinity of the Ca^{2+}-ATPase for Ca^{2+}.[12] The effect is exerted by organic anions, since it can not be reproduced with SO_4^{2-}, CO_3^{2-}, CH_3-COO^-, or NO_3^-. The effect of poly(L-aspartic acid) can be blocked with 30 μM trifluoperazine.

A small protein of apparent molecular weight 19,000 present in the cytosol of human red blood cells was purified to near homogeneity by Wuthrich.[13] When added to a suspension of disrupted red blood cell membranes in concentrations up to 1.5 μg/mℓ, it causes inhibition of Ca^{2+}-ATPase activity by decreasing Ca^{2+}-apparent affinity in more than one order of magnitude under conditions in which maximum velocity remains constant.

2. The EGTA Effect

Apart from its implications in the knowledge of factors determining the binding of Ca^{2+} at the transport site, the so-called "EGTA effect" is of great experimental relevance, since most of the activities of the Ca^{2+} pump are measured in media containing EGTA.

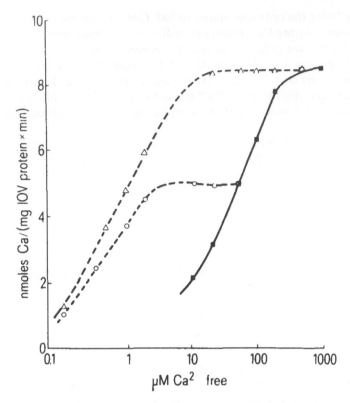

FIGURE 2.   The Ca²⁺ uptake by IOVs of red cell membranes as a function of Ca²⁺ concentration in the presence of no chelator (■); CaEGTA buffer containing 500 μM total calcium (Δ); and CaEGTA buffer containing 50 μM total calcium (O). (From Sarkadi, B., Schubert, A., and Gárdos, G., *Experientia*, 35, 1045, 1979. With permission.)

After Schatzmann observed that the K_{Ca} value during ATP hydrolysis by resealed ghosts was near 100-fold lower when CaEGTA instead of CaCl₂ was used,[8] Sarkadi et al.[14] measured the rate of Ca²⁺ uptake by IOVs from human red cells as a function of the concentration of Ca²⁺ in media with and without EGTA (Figure 2). They found that in Ca medium, K_{Ca} is 40 to 50 μM while in CaEGTA buffer, K_{Ca} is between 0.5 and 0.7 μM, that is, about 100-fold less than in the unbuffered solution. Exogenous calmodulin increases Jm and lowers from 40 to 15 μM the value of K_{Ca} in media without EGTA, but has no effect on the K_{Ca} in CaEGTA buffer. These results were confirmed by others.[15] Sarkadi et al.[14] explained their results in the following way. The Ca²⁺ pump from red blood cells has two different sites for calcium: site 1 and site 2. Both must be occupied for transport to occur. Site 1 has a K_{Ca} near 1 μM while site 2 has a K_{Ca} of about 10 μM in the presence of calmodulin and of about 50 μM in the absence of calmodulin. If we assume that site 2 does not discriminate between Ca²⁺ and CaEGTA and that both ligands are equally effective in promoting Ca²⁺ transport, the results are the expected, because the amount of calcium that will count for the activation is the total (chelated plus free) instead of the free Ca²⁺. Obviously, according to this view, Ca²⁺ binds to the Ca²⁺ pump with low affinity and all the effects of calmodulin are exerted on site 2.

However, as an alternative explanation for the observed effects, Schatzmann[16] thinks that it is possible that the access of Ca²⁺ to its binding site is facilitated in the CaEGTA complex; for instance, if the site were screened by a barrier of positive charges. If the rate of transport were higher than the rate or access of Ca²⁺ to its binding site, the CaEGTA complex will appear to increase the apparent affinity of the pump for Ca^{2+}. On this ground, Ca^{2+} binds to the Ca^{2+} pump with high affinity and the role of calmodulin would be to facilitate the access of Ca^{2+} to its binding site.[16]

As mentioned above, the EGTA effect was first observed on Ca^{2+}-ATPase activity in resealed ghosts from human red blood cells by Schatzmann in 1973.[8] Later on, Schatzmann (personal communication) and Al-Jobore and Roufogalis[17] confirmed it in isolated membranes from human red blood cells measuring Ca^{2+}-ATPase activity. Schatzmann found that the effect of EGTA on the Ca^{2+} affinity of the Ca^{2+}-ATPase is lost if the membranes are freeze-thawed (personal communication) as if, according to his theory, the treatment destroys the barrier of positive charges around the Ca^{2+} site. The need of the intact membrane structure for the EGTA effect has not been confirmed by Al-Jobore and Roufogalis,[17] because they have found that EGTA also increases the apparent affinity for Ca^{2+} during Ca^{2+}-ATPase activity of isolated membranes that have been solubilized with Triton X-100®, indicating that the effect does not depend on an intact membrane.

In a recent publication, Kotagal et al.[18] have extended the results on the "EGTA effect" to the Ca^{2+}-ATPase activity of pancreatic islet cells from the rat. They reported that at constant micromolar levels of Ca^{2+}, an increase in the concentration of EGTA from 0.2 to 2 mM lowers the K_{Ca} of the Ca^{2+}-ATPase from 36.9 to 0.5 μM while Vm rises from 0.008 to 0.075 μmol/mg protein/min. Tested on Ca^{2+} uptake by a plasma membrane-enriched fraction from the islet cells, the effects of EGTA were essentially similar to those on Ca^{2+}-ATPase activity. Furthermore, the EGTA effect appears to be specific for the Ca^{2+} pump of plasma membrane since EGTA does not change the kinetic parameters of Ca^{2+} uptake by membrane vesicles of endoplasmic reticulum from the same cells.[18] Kotagal et al.[18] also reported that the Ca^{2+}-ATPase from pancreatic cells maximally activated by EGTA is not further stimulated by calmodulin, whereas the stimulation by calmodulin increases as the concentration of EGTA in the medium decreases. Apart from describing a property of the Ca^{2+} pump from islet cells, these findings are important, because they suggest that the EGTA effect is not a property of the Ca^{2+} pump from human red blood cells, but rather a property of the Ca^{2+} pump from plasma membranes. Furthermore, the EGTA effect on Ca^{2+} affinity should be taken into account when testing the effects of calmodulin or calmodulin-like activators on Ca^{2+}-dependent activities of the Ca^{2+} pump from plasma membranes, because they could be hindered by the chelator.

C. The Number of Ca^{2+} Sites

The number of high-affinity sites for Ca^{2+}, present on each pump unit, is not known with certainty yet. The question is related to the number of Ca^{2+} that has to combine with the ATPase system to start the reaction cycle.

Ferreira and Lew[7] measured cytoplasmic calcium buffering, which relates the steady-state internal calcium concentration to the external Ca^{2+} concentration, and the rate of transport in intact red blood cells whose calcium content was varied with A23187, finding that only the square of the pump saturation related linearly with the concentration of internal Ca^{2+} as if, under the conditions of the experiment, Ca^{2+} binds at two sites of similar affinity. Additional evidence favoring this view was brought by experiments showing that in Ca^{2+}-loaded intact red blood cells, the Ca^{2+} pump rate changes according with the square of the concentration of intracellular Ca^{2+}.[19]

If the Ca^{2+} pump has two sites for Ca^{2+}, both of which must be occupied for the initiation of the transport cycle, the curve relating the functioning of the pump with Ca^{2+} concentration should be sigmoidal rather than hyperbolic. In Chapter 6, Section II.A, we have mentioned experimental evidence showing that the Hill coefficient of the Ca^{2+} activation curve of the Ca^{2+}-ATPase associated with calmodulin is significantly higher than one. If these results reflect the behavior of the enzyme moiety, it has to be concluded that during Ca^{2+} transport, Ca^{2+} combines with the pump at two or more high-affinity sites. However, care must be taken in interpreting these results for two reasons: (1) as stated by Schatzmann and Roelofsen,[20] when the experiments are performed at very low Ca^{2+} levels in media containing EGTA, the conclusion depends entirely on the assumption that the calculated concentration of Ca^{2+} is equal to its actual concentration in the assay medium, and (2) since calmodulin increases the affinity of the Ca^{2+}-ATPase for Ca^{2+} and binds to the Ca^{2+}-ATPase in a Ca^{2+}-dependent way, it may cause the appearance of sigmoidal kinetics even if the Ca^{2+} pump had only one site for Ca^{2+}.

If two or more Ca^{2+}-binding sites have to be occupied for Ca^{2+} transport, it remains still to be cleared whether one or all the bound Ca^{2+} is transported, because the possibility exists that, although necessary for activation, some of the calcium ions bound to the Ca^{2+} pump are not translocated. This question could be solved by knowing the relationship between the number of Ca^{2+} transported per ATP molecule hydrolyzed. Unfortunately, as will be seen in Section II.B of this chapter, the value of this stoichiometric relationship is still uncertain.

D. Dependence on pH

The rate of Ca^{2+} transport measured in in vitro experiments reaches a maximum at a pH between 7.5 and 8. In Ca^{2+}-loaded intact human red blood cells, the rate of Ca^{2+} transport is maximum when the pH of the suspending medium is 7.7.[21] Lee and Shin have reported that the transport of Ca^{2+} from resealed ghosts exhibits a constant rate from pH 6 to 9.5.[22] We have found that the transport of Ca^{2+} from resealed ghosts is optimum at pH 7.5 and drops by 35% at pH 8. In studying the pH dependence of active transport in either intact cells or resealed ghosts, only the pH of the suspending medium is known. Hence, no precise information on the requirements of pH at the inner surface of the membrane can be collected with these preparations. Such information can be made available from IOVs. There is no complete agreement among the authors who have studied the effect of pH on Ca^{2+} transport in this preparation. The rate of Ca^{2+} uptake has been reported to be maximum at pH 7.4 to 7.6[23] as well as at pH 6.9 to 7.3.[10] Furthermore, according to Waisman et al.,[15] the uptake of Ca^{2+} by IOVs almost doubles in going from pH 7.0 to pH 8.2, a behavior that seems to depend on calmodulin, since in its presence the rate of uptake remains constant between pH 7.0 and 7.8 and falls at pH 8.2.

The pH dependence of Ca^{2+} transport has also been studied in membrane preparation of cells other than the red blood cell. The ATP-driven Ca^{2+} uptake in vesicles from heart sarcolemma plasma membrane is optimum at pH 7.4 and drops by about 80% and 20% at pH 6.3 and pH 8.0, respectively.[24] DiPolo and Beaugé[25] took advantage of the squid axon and in a dialyzed preparation, studied the dependencies of active Ca^{2+} transport on internal and external pH and found them to be different (Figure 3). At pH 7.3 in the internal medium, the Ca^{2+} efflux is maximum and remains constant up to pH 8.5. Lowering the internal pH to 6.0 results in a marked decrease in the rate of Ca^{2+} efflux which lowers to about 10% of the maximum. The transport of Ca^{2+} through the Ca^{2+} pump of the squid axon is optimal at pH 7.0 to 7.6 in the external medium. Alkalinization from pH 7.6 to 9.0 causes a 35% inhibition of the activity. Inhibition by pH does not express any damage to the transport system, since DiPolo and Beaugé demonstrated that it can be reversed by bringing the pH to its initial value.[25]

The data show that the Ca^{2+}-ATPase activity depends on pH in a way that resembles that of active transport. Wins and Schoffeniels[26] reported that the Sr^{2+}-dependent activity of human red blood cell membranes vs. pH gives a bell-shaped curve with a maximum near pH 7.0. The activities at pH 5.0 and pH 9.0 are less than 5% of those

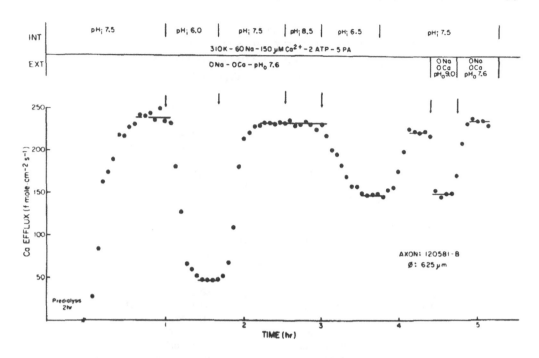

FIGURE 3. The effect of internal and external pH on the rate of Ca²⁺ pumping by the dialyzed squid axon. (From DiPolo, R. and Beaugé, L., *Biochim. Biophys. Acta,* 688, 237, 1984. With permission.)

measured at pH 7.0. The Ca^{2+}-ATPase activity of rat adipocyte plasma membranes is maximum at pH 7.8, and very little change in activity can be observed up to pH 8.4, while at pH 6.5, the activity of the enzyme is less than 40% the maximum.[27] The Ca^{2+}-ATPase activity that belongs to the Ca^{2+} pump from bone cell membranes exhibits a pH optimum of 7.2 and marked drops in activity can be observed at pH 8.4 and 6.0.[28] It has also been reported that optimum pH for ATPase activity may depend on calmodulin, since Ca^{2+}-dependent ATP hydrolysis by calmodulin-deficient human red blood cell membranes is maximal at pH 7.4, while for the same preparation saturated with calmodulin, it is maximal at pH 7.15.[29]

From the data given above, it is clear that concerning the pH dependence of the Ca^{2+} pump, a clear picture has not yet emerged. One cause for this may be that some of the experiments have been performed using EGTA, whose characteristics as a chelator could change at pH below 7.5.[30] Nevertheless, as has been mentioned previously, there seems to be agreement in that optimum pH for Ca^{2+} transport and Ca^{2+}-ATPase is around 7.5 to 8.0 and that exposing the internal surface of the membrane to pH below 7.0 lowers the activity of the pump. These effects of pH could be mainly due to two causes: (1) changes in the ionization of the ligands needed for Ca^{2+} transport and/or of the charged groups in or near the sites of binding of the ligands or (2) the imposition of a H^+ concentration gradient against the transport of H^+ that, as mentioned in Chapter 6, Section I.E, may take place during transport of Ca^{2+}. There is no experimental evidence to decide between these two alternatives.

E. Dependence on Temperature

Table 1 shows values of apparent activation energy for the active transport of Ca^{2+} in various preparations from human red blood cell membranes. Above 20°C, values are around 60 kJ/mℓ. Kratje et al.[9] and Mollaman and Pleasure[10] reported that the Arrhenius plot of Ca^{2+} efflux from resealed ghosts and Ca^{2+} uptake by IOVs shows a break in the vicinities of 20°C. This could tentatively be attributed to phase transition

Table 1
THE TEMPERATURE DEPENDENCE OF Ca²⁺ TRANSPORT ACROSS PLASMA MEMBRANE FROM HUMAN RED BLOOD CELLS

Preparation	Activity measured	Q_{10}	Apparent activation energy (kJ/mol)	Ref.
Intact cells	Initial rate of Ca²⁺ efflux	3.6ᵃ (6—31)	63.5	21
Resealed ghosts	Half-time of Ca²⁺ efflux	3.16 (20—40)	56.8	22
Resealed ghosts	Initial rate of Ca²⁺ efflux	3.5 (14—28)	104.5	1
Resealed ghosts	Initial rate of Ca²⁺ efflux	5.5 (10—20) 2.2 (20—37)	117 60.2	9
IOVs	Rate of Ca²⁺ uptake Plus exogenous calmodulin		80.3 (12—39) Not significantly different to the control	23
IOVs	Initial rate of Ca²⁺ uptake	4.27 (15—25) 2.14 (30—40)	142.5 59.7	10

Note: The figures in parentheses are the temperature range in °C at which the activity was measured. Where two values are given, the Arrhenius plot of the data showed a break point.

ᵃ Calculated from their data.

of the lipids in the membrane. It seems worth commenting here that the Arrhenius plot for Ca²⁺ transport and Ca²⁺-ATPase activity in intact sarcoplasmic reticulum vesicles shows a break at about 20°C, and the apparent activation energies below and above this temperature are 92.4 and 69.4 kJ/mol, respectively,[31] just about as observed in red blood cell preparations. The Arrhenius plots of sarcoplasmic reticulum vesicles in which most of the lipids have been replaced by detergents are strikingly similar to those of the natural membrane. This has been interpreted to mean that in sarcoplasmic reticulum, lipid-protein interactions are not responsible for the breaks seen in the Arrhenius plots and that it argues instead for temperature-dependent changes intrinsic to the enzyme reaction.[31] The possibility that the same holds for Ca²⁺ transport across the red blood cell membrane cannot be discarded.

Many of the authors cited in Table 1 have also reported that above 42°C, Ca²⁺ transport is very small or null and its coupling with Ca²⁺-dependent ATP hydrolysis lost. It is reasonable to attribute this temperature effect to an increase in the leakiness of the membrane to Ca²⁺ at that temperature. Again, as in red blood cells, no transport of Ca²⁺ can be detected in sarcoplasmic reticulum vesicles above 42°C.[31]

II. THE COUPLING BETWEEN Ca²⁺ TRANSPORT AND ATP HYDROLYSIS

A. Energetics of Ca²⁺ Transport

The reaction of Ca²⁺ transport linked to ATP hydrolysis across a plasma membrane can be written:

$$nCa_i^{2+} + ATP \rightleftharpoons nCa_0^{2+} + ADP + Pi \qquad (1)$$

where the subscript "i" denotes intracellular and "o" extracellular Ca²⁺. The factor "n", the stoichiometric coefficient, is a positive integer denoting the number of Ca²⁺ transported per ATP molecule hydrolyzed. Its value can be determined experimentally. For Reaction 1 to proceed from left to right, the sum of the Gibbs energy changes of the hydrolysis of ATP and the transport of Ca²⁺ must be negative. This can be determined by means of the equation:

$$\Delta G_t = \Delta G^{\circ\prime} + RT \ln \frac{[ADP]\,[Pi]}{[ATP]} + nRT \ln \frac{[Ca_0^{2+}]}{[Ca_i^{2+}]} + n2FE$$

$$\underset{\longleftarrow \; a \; \longrightarrow}{} \quad \underset{\longleftarrow \; b \; \longrightarrow}{} \underset{\longleftarrow c \longrightarrow}{} \qquad (2)$$

where $\Delta G^{\circ\prime}$ = −30.55 kJ mol⁻¹; R = 8.31 J mol⁻¹ K⁻¹; F = 96,500 coul geq⁻¹ T = 310 K; n = the stoichiometric coefficient. For a normal eukaryotic cell, cytosolic ATP, ADP, and Pi are near 1.5×10^{-3} M, 0.3×10^{-3} M, and 0.3×10^{-3} M, respectively. The concentration of cytosolic calcium ion $[Ca_i^{2+}]$ = 10^{-7} M and the concentration of extracellular calcium ion $[Ca_0^{2+}]$ can be taken as 10^{-3} M. E represents the membrane potential which varies from cell to cell between 0.01 V to 0.1 V, negative inside. For the concentrations mentioned above, the value of "a" in Equation 2 is −55.6 kJ/mol and "b", which represents the osmotic work, can be either 23.7 kJ/mol for n = 1 or 47.4 kJ/mol for n = 2. The electric work is represented by "c" and its value can go from 1.93 kJ when n = 1 and E = 0.01 V to 38.6 kJ when n = 2 and E = 0.1 V.

For a cell like the erythrocyte where the membrane potential is about 0.01 V, the numerical value of the electric work is small in comparison with that of the osmotic work done during the transport reaction. The value of ΔG_t will be negative even if n = 2, and there will be no thermodynamic restrictions for Reaction 1 to take place from left to right. The same does not apply to all cells, because for a membrane potential of 0.1 V when n = 2, ΔG_t will be positive and the transport reaction will not proceed from left to right.

B. The Stoichiometry of Ca²⁺ Transport

Table 2 summarizes the data in the literature on the radio between the number of Ca²⁺ transported and ATP molecules hydrolyzed. Most of the experiments have been performed in red blood cell preparations. Values range from 0.86 to 2.1. The data also show that in tissues other than the red blood cell values close to 1 as well as 2 have been reported. These experimental results together with the energetics of the transport reaction indicate that the value of the stoichiometric coefficient is either 1 or 2. Experimental values of the stoichiometric coefficient far from 1 or 2 could also be attributed to: (1) membrane preparations not tight to Ca²⁺ which allow the cation to leak back to the medium from which it was transported, (2) partial uncoupling of the ATPase reaction, and (3) factors that shift the apparent stoichiometry.

Procedures used for measuring the stoichiometry of the transport reaction have con-

Table 2

THE NUMBER OF Ca²⁺ TRANSPORTED PER ATP MOLECULE HYDROLYZED BY THE Ca²⁺ PUMP FROM VARIOUS MEMBRANE PREPARATIONS

Type of preparation	Ca²⁺/ATP	Activities measured	Ref.
Resealed ghosts from HRBC	1.39	Downhill Ca²⁺ efflux and ATP hydrolysis	8
Resealed ghosts from HRBC	1.27	Uphill Ca²⁺ efflux and ATP hydrolysis	8
Resealed ghosts from HRBC	2.02	LaCl₃-sensitive, Ca²⁺ efflux and ATP hydrolysis	32
Intact HRBC	1.96	LaCl₃-sensitive, Ca²⁺ efflux and ATP hydrolysis in the presence of iodoacetamide and tetrathionate	21
IOVs from HRBC	1.5	Ca²⁺ uptake and ATP hydrolysis	33
	2.0	Ca²⁺ uptake and ATP hydrolysis minus ATP hydrolysis due to unsealed vesicles	
Resealed ghosts from HRBC	0.86	Ca²⁺ efflux and ATP hydrolysis; the value is the slope of the straight line relating the two activities at various concentrations of La³⁺; Ca²⁺ efflux measured with a Ca²⁺ selective electrode	34
	0.86	The same but Ca²⁺ efflux measured by atomic absorption spectroscopy	
Resealed ghosts from HRBC	0.93	Ca²⁺ efflux and ATP hydrolysis at 10 μM ATP	35
	1.08	The same at 500 μM ATP	
Purified Ca²⁺-ATPase from pig erythrocyte reconstituted in asolectin liposomes	0.79	Ca²⁺ uptake and (³²P)ATP hydrolysis	36
	0.77	Ca²⁺ uptake and (³²P)ATP hydrolysis in the presence of CM	
IOVs from HRBC	1.6	Ca²⁺ uptake and ATP hydrolysis	15
	0.94	Ca²⁺ uptake and ATP hydrolysis in the presence of CM	
Basolateral plasma membrane vesicles from rat duodenum	1.08	Ca²⁺ uptake and ATP hydrolysis assuming that half of the resealed vesicles are inside-out	37
Purified Ca²⁺-ATPase from HRBC incorporated in asolectin liposomes	0.94	Ca²⁺ uptake and (³²P) ATP hydrolysis	38
Purified Ca²⁺-ATPase of calf heart sarcolemma reconstituted in asolectin liposomes	Near 1.0	Ca²⁺ uptake and ATP hydrolysis estimated assuming 0.7 mol of H⁺ produced per ATP hydrolyzed	39
Plasma membrane vesicles from neutrophil leukocytes	2.1	Ca²⁺ uptake and ATP hydrolysis assuming that 33% are sealed and 26% leaky IOVs	40, 41
Basolateral plasma membrane vesicles from rat kidney cortex	0.7	Ca²⁺ uptake and ATP hydrolysis assuming that 12.5% of the membrane vesicles are IOVs	42

Note: HRBC = human red blood cell.

sisted in comparing the rate of transport with the rate of ATP hydrolysis measured under identical conditions in aliquots of the same preparation.

Before purified preparations of the Ca²⁺ pump were available, Quist and Roufogalis[32] introduced the idea of comparing the fraction of Ca²⁺ transport and of Ca²⁺-ATPase activity which are sensitive to a given inhibitor applied under identical conditions. They used LaCl₃ (see Chapter 11) and found that under conditions that Ca²⁺ transport ceased, about 50% of the Ca²⁺-ATPase activity in resealed ghosts was still present. The obvious conclusion is that in the preparation of red blood cell membranes used, only 50% of the total ATP hydrolysis observed in medium with $Mg^{2+} + Ca^{2+}$ is coupled to Ca^{2+} transport. The authors concluded that two Ca^{2+} are transported per molecule of ATP hydrolyzed and hence that in resealed ghosts from red blood cells, half of the Ca^{2+}-dependent ATP hydrolysis is catalyzed by an ATPase system other than the Ca^{2+} pump.

The interference of ATP-hydrolyzing systems other than the Ca^{2+}-ATPase during measurements of the stoichiometry of Ca^{2+} transport can be ignored if purified preparations of the Ca^{2+}-ATPase incorporated into liposomes are used. In preparations of this sort obtained from pig[36] and human[38] red blood cells and heart sarcolemma,[39] the stoichiometry of the transport reaction was found to be close to 1 and independent of calmodulin (Table 1). The experiment of Muallem and Karlish (Table 1) in resealed ghosts from red blood cells demonstrates that the stoichiometry is independent of the degree of occupation of the sites for ATP (Figure 5, Chapter 4), since it remains close to 1 for the whole range of ATP concentrations tested which go from 10 to 500 μM.

In contrast with the lack of effect of ATP concentration, there are reports showing that the ratio Ca^{2+} transported/ATP hydrolyzed is influenced by Ca^{2+} and Mg^{2+}. Sarkadi[43] reported experiments in IOVs of human red blood cells showing that as a function of calcium concentration in the medium, the ratio goes from 0.3 at about 50 μM calcium to 2 at 700 μM calcium. Akyempon and Roufogalis[44] showed that in IOVs from red blood cells, the apparent stoichiometry depends on Ca^{2+} and Mg^{2+} and is independent of calmodulin. According to their results, at 6.4 mM $MgCl_2$, two Ca^{2+} are transported per ATP hydrolyzed at Ca^{2+} concentrations below 24 μM and only one above 50 μM Ca^{2+}. If $MgCl_2$ is reduced from 6.4 mM to 1 mM, the stoichiometry remains 2 over all the range of Ca^{2+} concentrations tested. Furthermore, Waisman et al.[45] studied the stimulation of Ca^{2+}-ATPase and Ca^{2+} transport by anions like chloride, sulfate, acetate, and phosphate in IOVs, finding that a given anion does not stimulate both activities to a similar degree so that the apparent stoichiometry varied from 0.25 in the case of chloride to 0.87 in the case of phosphate. The apparent dependence of the number of Ca^{2+} transported per ATP hydrolyzed during the transport reaction with the assay conditions could explain, at least in part, the noninteger values and the discrepancies among the different reports shown in Table 1. Nevertheless, whether the stoichiometry of the Ca^{2+} pump from plasma membranes is 1 or 2 still remains an open question. It is generally accepted that in sarcoplasmic reticulum, calcium movement and ATP hydrolysis are stoichiometrically connected by a coupling ratio of two.[46] Here again, it has been reported that depending upon substrate concentration and pH, the ratio of Ca^{2+} transported over ATP hydrolyzed by the Ca^{2+} pump from sarcoplasmic reticulum can be anything from no calcium transported during ATP hydrolysis to 2.0 Ca^{2+} transported per ATP hydrolyzed.[47]

C. Reversal of the Ca^{2+} Pump

If the hydrolysis of ATP plus the vectorial reaction of Ca^{2+} translocation are catalyzed by the single molecule or molecular complex that makes up the Ca^{2+} pump to energize Ca^{2+} outflow, Ca^{2+} entry through the Ca^{2+} pump should energize ATP synthesis form ADP and Pi. Equation 2 predicts that when the Gibbs energy change necessary for the extrusion of "n" calcium ions is made higher than that of the hydrolysis of ATP, the overall transport reaction will reverse, proceeding from right to left with net synthesis of ATP, the energy coming from the Ca^{2+} concentration gradient across the membrane. Hence, for synthesis of ATP in a cell to occur by reversal of the Ca^{2+} pump, in the absence of membrane potential, $[Ca^{2+}_o]$ must be higher relative to $[Ca^{2+}_i]$. A low ATP and a high ADP and Pi concentration will help. In all reported experiments made with that purpose, Ca^{2+}_i and ATP were made as low and Pi and Ca^{2+}_o as high as possible.

For instance, we know by experience collected in our laboratory that if human red blood cells are starved and loaded with orthophosphate, cells containing $0.13 \times 10^{-3} M$ ATP, $0.23 \times 10^{-3} M$ ADP, and $22.4 \times 10^{-3} M$ Pi can be obtained.[48] If these cells were suspended in medium containing $2 \times 10^{-3} M$ Ca²⁺, assuming that their intracellular Ca²⁺ concentration was $10^{-6} M$, the total Gibbs energy change calculated according to Equation 2 would be near −17 kJ/mol (negative) for n = 1 or near 4 kJ/mol (positive) for n = 2. Hence if n were 2, net synthesis of ATP is expected under these experimental conditions and if n were 1, hydrolysis rather than net synthesis of ATP is to be expected. Whatever value for n is valid, the backward component in these starved cells should be larger than in fresh cells so as to reach measurable levels.

Ferreira and Lew[49] reported that if isotonically resealed ghosts from human red blood cells containing a 3 mM EGTA are exposed to extracellular calcium at concentrations that produce a measurable entry of the cation into the cell, an incorporation of ^{32}Pi into ATP is detectable as if Ca²⁺ entered the cell through the Ca²⁺ pump.

We have measured the incorporation of (^{32}P)Pi into ADP to form ATP in starved and phosphate-loaded red blood cells with the composition mentioned above, suspended in media with and without 2 mM Ca²⁺.[48] There was some incorporation, whatever the medium was, but in every experiment there was an extra incorporation when the cells were in the calcium medium. The extra incorporation was abolished by either the ionophore A23187, which collapses the Ca²⁺ concentration gradient across the membrane, or by LaCl₃, which blocks the Ca²⁺ pump. The cells in which the extra incorporation was detected also showed an uptake of Ca²⁺ not apparent in fresh cells. These results are the expected if the incorporation of Pi into ATP were catalyzed by the Ca²⁺ pump using the energy of the Ca²⁺ concentration gradient. To demonstrate net ATP synthesis by the Ca²⁺ pump, Wuthrich et al.[50] took advantage of the IOVs which allow the composition of the medium in contact with the internal surface of the membrane in the intact cell to be fixed with precision. For that purpose, the authors Ca²⁺ loaded the vesicles with up to about 20 mM Ca²⁺ by feeding the pump with ATP. Then, the vesicles were washed and suspended in a Ca²⁺-free medium containing no ATP, 2 mM ADP and 10 mM Pi. With ATP $2.5 \times 10^{-6} M$ and assuming zero membrane potential, the total Gibbs energy change can be calculated to be 12 kJ/mol for n = 1 or 50 kJ/mol for n = 2 (both positive). Under these conditions, the vesicles produced systematically an excess of ATP compared to controls incubated in medium containing A23187.

These series of results confirm that the Ca²⁺ pump of plasma membranes can be forced to run in a reverse fashion utilizing the energy derived from the Ca²⁺ concentration gradient to synthetize ATP.

REFERENCES

1. Schatzmann, H. J. and Vincenzi, F. F., Calcium movements across the membrane of human red cells, *J. Physiol.*, 201, 369, 1969.
2. Olson, J. E. and Cazort, R. J., Active calcium and strontium transport in human erythrocyte ghosts, *J. Gen. Physiol.*, 53, 311, 1969.
3. Sarkadi, B., Szebeni, J., and Gárdos, G., Effects of calcium on cation transport processes in inside-out red cell membrane vesicles, in *Membrane Transport in Erythrocytes*, Lassen, U. V., Ussing, H. H., and Wieth, J. D., Eds., Munksgaards, Copenhagen, 1980, 220.
4. Graf, E., Verma, A. K., Gorski, J. P., Lopaschuk, G., Niggli, V., Zurini, M., Carafoli, E., and Penniston, J. T., Molecular properties of calcium-pumping ATPase from human erythrocytes, *Biochemistry*, 21, 4511, 1982.
5. Pfleger, H. and Wolf, H. U., Activation of membrane-bound high-affinity calcium ion-sensitive adenosine triphosphatase of human erythrocytes by bivalent metal ions, *Biochem. J.*, 147, 359, 1975.

6. Schatzmann, H. J., Active calcium transport and Ca²⁺-activated ATPase in human red cells, *Current Topics in Membranes and Transport*, 6, 125, 1975.

7. Ferreira, H. G. and Lew, V. L., Use of ionophore A23187 to measure cytoplasmic Ca buffering and activation of the Ca pump by internal Ca, *Nature*, 259, 47, 1976.

8. Schatzmann, H. J., Dependence on calcium concentration and stoichiometry of the calcium pump in human red cells, *J. Physiol.*, 235, 551, 1973.

9. Kratje, R. B., Garrahan, P. J., and Rega, A. F., The effects of alkali metal ions on active Ca²⁺ transport in reconstituted ghosts from human red cells, *Biochim. Biophys. Acta*, 731, 40, 1983.

10. Mollman, J. E. and Pleasure, D. E., Calcium transport in human inside-out erythrocyte vesicles, *J. Biol. Chem.*, 255, 569, 1980.

11. Roufogalis, B. D., Akyempon, Ch. K., Al-Jobore, A., and Minocherhomjee, A. M., Regulation of the Ca²⁺ pump of the erythrocyte membrane, *Ann. N.Y. Acad. Sci.*, 402, 349, 1982.

12. Minocherhomjee, A., Al-Jobore, A., and Roufogalis, B. D., Modulation of the calcium-transport ATPase in human erythrocytes by anions, *Biochim. Biophys. Acta*, 690, 8, 1982.

13. Wuthrich, A., Isolation from homeolysate of a protein across inhibitor of the red cell Ca²⁺-pump ATPase. Its action on the kinetics of the enzyme, *Cell Calcium*, 3, 201, 1982.

14. Sarkadi, B., Schubert, A., and Gárdos, G., Effects of calcium-EGTA buffers on active calcium transport in inside-out red cell membrane vesicles, *Experientia*, 35, 1045, 1979.

15. Waisman, D. M., Gimble, J. M., Goodman, D. B. P., and Rasmussen, H., Studies of the Ca²⁺ transport mechanism of human erythrocyte inside-out plasma membrane vesicle. I. Regulation of the Ca²⁺ pump by calmodulin, *J. Biol. Chem.*, 256, 409, 1981.

16. Schatzmann, H. J., The plasma membrane calcium pump of erythrocytes and other animal cells, in *Membrane Transport of Calcium*, Carafoli, E., Ed., Academic Press, London, 1982, chap. 2.

17. Al-Jobore, A. and Roufogalis, B. D., Influence of EGTA on the apparent Ca²⁺ affinity of Mg²⁺-dependent, Ca²⁺-stimulated ATPase in the human erythrocyte membrane, *Biochim. Biophys. Acta*, 645, 1, 1981.

18. Kotagal, N., Colca, J. R., and McDaniel, M. L., Activation of an islet cell plasma membrane (Ca²⁺ - Mg²⁺)-ATPase by calmodulin and Ca-EGTA, *J. Biol. Chem.*, 258, 4808, 1983.

19. Lew, V. L., Tsien, R. Y., Miner, C., and Bookchin, R. M., Physiological [Ca²⁺]ᵢ level and pump-leak turnover in intact red cells measured using an incorporated Ca chelator, *Nature*, 298, 478, 1982.

20. Schatzmann, H. J. and Roelofsen, B., Some aspects of the Ca pump in human red blood cells, in *Biochemistry of Membrane Transport*, Semenza, G. and Carafoli, E., Eds., Springer-Verlag, Berlin, 389, 1977.

21. Sarkadi, B., Szász, J., Gerlóczy, A., and Gárdos, G., Transport parameters and stoichiometry of active calcium ion extrusion in intact human red cells, *Biochim. Biophys. Acta*, 464, 93, 1977.

22. Lee, K. S. and Shin, B. C., Studies on the active transport of calcium ion human red cells, *J. Gen. Physiol.*, 54, 713, 1969.

23. Sarkadi, B., Szász, J., and Gárdos, G., Characteristics and regulation of active calcium transport in inside-out red cell membrane vesicles, *Biochim. Biophys. Acta*, 598, 326, 1980.

24. Caroni, P. and Carafoli, E., The Ca²⁺ pumping ATPase of heart sarcolemma, *J. Biol. Chem.*, 256, 3263, 1981.

25. DiPolo, R. and Beaugé, L., The effect of pH on Ca²⁺ extrusion mechanisms in dialyzed squid axons, *Biochim. Biophys. Acta*, 688, 237, 1982.

26. Wins, P. and Schoffeniels, E., Studies on red-cell ghost ATPase systems: properties of a (Mg²⁺ + Ca²⁺)-dependent ATPase, *Biochim. Biophys. Acta*, 120, 341, 1966.

27. Pershadsingh, H. A. and McDonald, J. M., A high affinity calcium-stimulated magnesium-dependent adenosine triphosphatase in red adipocyte plasma membranes, *J. Biol. Chem.*, 255, 4087, 1980.

28. Shen, V., Kohler, G., and Peck, W. A., A high affinity, calmodulin responsive (Ca²⁺ - Mg²⁺)-ATPase in isolated bone cells, *Biochim. Biophys. Acta*, 727, 230, 1983.

29. Scharf, O., Regulation of (Ca²⁺, Mg²⁺)-ATPase in human erythrocytes dependent on calcium and calmodulin, *Acta Biol. Med. Germ.*, 40, 457, 1981.

30. Wolf, H. V., Divalent metal ion buffers with low pH-sensitivity, *Experientia*, 29, 241, 1973.

31. de Meis, L., The sarcoplasmic reticulum. Transport and energy transduction, in *Transport in the Life Sciences*, Vol. 2., Bittar, E. E., Ed., John Wiley & Sons, New York, 1981.

32. Quist, E. E. and Roufogalis, B. D., Determinations of the stoichiometry of the calcium pump in human erythrocytes using lanthanum as a selective inhibitor, *FEBS Lett.*, 50, 135, 1975.

33. Quist, E. E. and Roufogalis, B. D., Association of (Ca + Mg)-ATPase activity with ATP-dependent Ca uptake in vesicles prepared from human erythrocytes, *J. Supramol. Struct.*, 6, 375, 1977.

34. Larsen, F. L., Hinds, T. R., and Vincenzi, F. F., On the red blood cells Ca²⁺ pump: an estimate of stoichiometry, *J. Membrane Biol.*, 41, 361, 1978.

35. Muallem, S. and Karlish, S. J. D., Is the red cell calcium pump regulated by ATP?, *Nature*, 277, 238, 1979.

36. Haaker, H. and Racker, E., Purification and reconstitution of the Ca²⁺-ATPase from plasma membranes of pig erythrocytes, *J. Biol. Chem.*, 254, 6598, 1979.

37. Ghijsen, W. E. S. M., De Jong, M. D., and Van Os, C. H., ATP-dependent calcium transport and its correlation with Ca²⁺-ATPase activity in basolateral plasma membranes of rat duodenum, *Biochim. Biophys. Acta,* 689, 327, 1982.

38. Clark, A. and Carafoli, E., The stoichiometry of the Ca²⁺-pumping ATPase of erythrocytes, *Cell Calcium,* 4, 83, 1983.

39. Caroni, P., Zurini, M., Clark, A., and Carafoli, E., Further characterization and reconstitution of the purified Ca²⁺-pumping ATPase of heart sarcolemma, *J. Biol. Chem.,* 258, 7305, 1983.

40. Ochs, D. L. and Reed, P. W., ATP-dependent calcium transport in plasma membrane vesicles from neutrophil leukocytes, *J. Biol. Chem.,* 258, 10116, 1983.

41. Ochs, D. L. and Reed, P. W., Ca²⁺-stimulated, Mg²⁺-dependent ATPase activity in neutrophil plasma membrane vesicles. Coupling to Ca²⁺ transport, *J. Biol. Chem.,* 259, 102, 1984.

42. Van Heeswijk, M. P. E., Geertsen, J. A. M., and Van Os, C. H., Kinetic properties of the ATP-dependent Ca²⁺ pump and the Na⁺/Ca²⁺ exchange system in basolateral membranes from rat kidney cortex, *J. Membr. Biol.,* 79, 19, 1984.

43. Sarkadi, B., Active calcium transport in human red cells, *Biochim. Biophys. Acta,* 604, 159, 1980.

44. Akyempon, C. K. and Roufogalis, B. D., The stoichiometry of the Ca²⁺ pump in human erythrocyte vesicles: modulation by Ca²⁺, Mg²⁺ and calmodulin, *Cell Calcium,* 3, 1, 1982.

45. Waisman, D. M., Gimble, J. M., Goodman, D. B. P., and Rasmussen, H., Studies of the Ca²⁺ transport mechanism of human erythrocyte inside-out plasma membrane vesicles, *J. Biol. Chem.,* 256, 420, 1981.

46. Hasselbach, W. and Wass, W., Energy coupling in sarcoplasmic reticulum Ca²⁺ transport: an overview, *Ann. N.Y. Acad. Sci.,* 402, 459, 1982.

47. Rossi, B., Leone, F., Gache, C., and Lazdunski, M., Pseudosubstrated of the sarcoplasmic Ca²⁺-ATPase as tools to study the coupling between substrate hydrolysis and Ca²⁺ transport, *J. Biol. Chem.,* 254, 2302, 1979.

48. Rossi, J. P. F. C., Garrahan, P. J., and Rega, A. F., Reversal of the calcium pump in human red cell, *J. Membr. Biol.,* 44, 37, 1978.

49. Gil Ferreira, H. and Lew, V. L., Ca transport and Ca pump reversal in human red blood cells, *J. Physiol.,.* 252, 86P, 1975.

50. Wuthrich, A., Schatzmann, H. J., and Romero, P., Net ATP synthesis by running the red cell calcium pump backwards, *Experientia,* 35, 1789, 1979.

Chapter 8

PARTIAL REACTIONS OF THE Ca²⁺-ATPase

P. J. Garrahan

I. THE ELEMENTARY STEPS OF ATP HYDROLYSIS

A. Introduction

Hydrolysis of ATP by the Ca²⁺-ATPase of the plasma membrane takes place following a sequence of elementary steps that involve cyclic conformational transitions and cyclic formation and hydrolysis of a phosphoenzyme. This mechanism is essentially similar to that of the other cation-transport ATPases.[1-4]

The participation of an acid-stable phosphoenzyme as an intermediate in the reaction of hydrolysis of ATP has been extremely useful in the characterization of the mechanism of all cation-transport ATPases. In fact, most of our present knowledge on the elementary steps of the reaction has come from studies on the time course of the phosphoenzyme under presteady-state conditions. These studies are possible, because the ATPase reaction can be interrupted by acid denaturation without affecting the level of the phosphoenzyme.

Studies on presteady-state enzyme kinetics at near physiological temperatures require the use of stop-flow techniques having time resolutions of the order of milliseconds. These procedures have been applied in several cation-transport ATPases (see for instance References 5 and 6), but not in the Ca²⁺-ATPase of plasma membranes. It is likely that the recent availability of solubilized and purified preparations of Ca²⁺-ATPase will change this picture in the near future. However, until this happens, all our information about the elementary steps of the Ca²⁺-ATPase comes from studies performed at temperatures around 0°C, temperatures at which elementary steps have time-courses of the order of seconds and can be measured by more conventional techniques. Caution therefore must be exerted in extrapolating results to more physiological temperatures.

In what follows, we will first analyze separately each of the known elementary steps of the Ca²⁺-ATPase reaction, and then we will try to show how these steps can be combined to yield a plausible reaction scheme for ATP hydrolysis by the system.

B. Phosphorylation

The catalysis of hydrolysis of ATP by the Ca²⁺-ATPase begins with the Ca²⁺-dependent transference of the γ-phosphate of ATP to the enzyme with the formation of a covalent phosphoenzyme and the release of ADP according to the reaction:

$$ATP + E \underset{}{\overset{Ca^{2+}}{\rightleftharpoons}} E{\sim}P + ADP \tag{1}$$

where E is the Ca²⁺-ATPase and E ∿ P the phosphoenzyme.

The first experimental evidence for the existence of Reaction 1 was provided in 1974 by Knauf et al.[7] These authors incubated red blood cell membranes in media containing ATP labeled with ³²P at its γ-phosphate and demonstrated that the quantity of ³²P incorporated into the protein fraction of the membranes increased when Ca²⁺ was present in the reaction media. The Ca²⁺-dependent phosphorylation took place into a single polypeptide chain of an apparent M_r of about 140,000, which was not phosphorylated

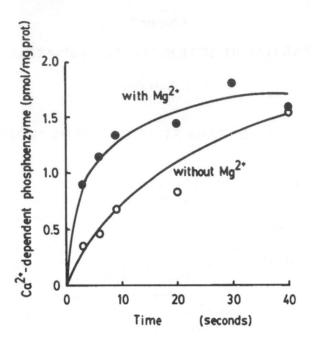

FIGURE 1.   Time course of phosphoenzyme formation
in human red blood cell membranes at 0°C in media containing 0 mM and 0.5 mM MgCl₂. The concentration of
(³²P-γ)ATP was 15 μM and that of Ca²⁺ 100 μM. The data
was taken from Figures 1A and 1B in Reference 21.

in the absence of Ca²⁺. These findings were confirmed and extended by Katz and Blonstein[8] and by ourselves.[9]

The identification of the phosphorylation reaction with an elementary step of ATP
hydrolysis was at first based on three indirect criteria, i.e., (1) the level of phosphoprotein varies with the concentrations of Ca²⁺ and of ATP in the same way as Ca²⁺-ATPase
activity;[8,9] (2) the phosphate in the protein undergoes rapid turnover decaying in a few
seconds at 0°C when unlabeled ATP is added or Ca²⁺ is removed,[8,9] and (3) the chemical properties of the protein-phosphate bond are similar to those of the phosphoenzyme of other cation transport ATPases.[8] The definitive proof was provided by the
demonstration that in solubilized and purified Ca²⁺-ATPase, ATP phosphorylates the
enzyme and that the phosphoenzyme formed has the same properties as the phosphoprotein of intact membranes.[10,12]

Although most of the properties of the phosphorylation reaction have been studied
in isolated red blood cell membranes, the reaction has also been detected in intact red
blood cells[13] and in cell membranes from several other tissues (see Chapter 4).

1. Kinetics of the Phosphorylation Reaction

Phosphorylation is the only elementary step of the ATPase reaction that is absolutely dependent on Ca²⁺. Hence, this reaction probably is responsible for the Ca²⁺-
dependence of the overall reaction, and it is likely that it participates in the first step
of active Ca²⁺ translocation.

A typical phosphorylation experiment of red blood cell membranes is shown in Figure 1. At 0°C the reaction reaches steady state with a half-time that is about 20 sec in
media without added Mg²⁺, and about 3 sec in media with nonlimiting concentrations
of Mg²⁺. The effect of Mg²⁺ seems to be exerted only in the rate of phosphorylation
since the steady-state level of phosphorylation is independent of added Mg²⁺.[14] As will

be discussed in more detail in Chapter 9, it is not yet possible to exclude the possibility that in the absence of added Mg^{2+}, endogenous Mg^{2+} either as a contaminant in the incubation media or as Mg^{2+} tightly bound to the enzyme is required for phosphorylation.

The steady-state level of phosphorylation increases with the concentration of Ca^{2+} and of ATP following hyperbolic kinetics[8,9] (Figure 2). In red blood cell membranes, K_{Ca} for the phosphorylation reaction is about 5 μM, a value close to the apparent affinity of the activating site for Ca^{2+} of the Ca^{2+}-ATPase in the same membrane preparation. Km is between 3 to 7 μM and hence similar to the Km of the high affinity component of the substrate curve of the Ca^{2+}-ATPase. It seems therefore reasonable to postulate that phosphorylation requires ATP at the high-affinity site and Ca^{2+} at the transport site of the Ca^{2+}-ATPase and that the high-affinity site for ATP is the catalytic site of the ATPase.

It is known that high-affinity sites for Ca^{2+} are detectable in the absence of ATP[15] and that the apparent affinity for ATP at the catalytic site is independent of Ca^{2+}.[16] Hence, it seems that to promote phosphorylation, ATP and Ca^{2+} bind in a random fashion at their sites in the enzyme. Since (1) there are no interactions between ATP and Ca^{2+} in their apparent affinities, (2) phosphorylation proceeds in the absence of added Mg^{2+}, and (3) phosphorylation reaches its maximum levels in the presence of a large excess of Ca^{2+} over Mg^{2+},[17] it seems likely that free ATP, CaATP, and MgATP are equally effective as substrates for the phosphorylation reaction. As mentioned in Chapter 6, Section III.D, a similar conclusion has been reached for the overall Ca^{2+}-ATPase activity.[18]

In red blood cell membranes, steady-state level of phosphorylation is about 3 pmol/mg of membrane protein. Values of the same order of magnitude have been measured in membranes from other cell types (see Chapter 4, Section II), which suggests that the surface density of Ca^{2+} pump units is about the same in all plasma membranes. The level of steady-state phosphorylation can be substantially increased by very high (10 mM concentrations of Ca^{2+},[17] by La^{3+},[13,18-20] or by K^+.[21] Hence, under the usual assay conditions, only a fraction of the ATPase molecules are in the phosphorylated state. In fact, it has been estimated by Luterbacher (cited by Schatzman[22]) that in media with physiological concentrations of ligands and at 0°C, less than 10% of the enzyme is in the phosphorylated form. This has to be taken into account, since it may lead to gross underestimations if the level of Ca^{2+}-dependent phosphorylation is used to estimate the number of Ca^{2+}-ATPase units in a given membrane.

2. Reversal of Phosphorylation

When phosphorylation is stopped by isotopic dilution of labeled ATP, the phosphoenzyme rapidly decays. The decay is accelerated by ADP. The effect of ADP depends on Ca^{2+},[23] suggesting that it expresses the ability of the phosphoenzyme to transfer its phosphate to ADP and regenerate ATP through the reversal of the phosphorylation reaction (Figure 3). Phosphorylation therefore seems to proceed with energy conservation.

3. Chemical Properties of the Phosphoenzyme

When phosphorylated membranes are denatured, dissolved in solutions containing sodium dodecyl sulfate (SDS), and submitted to electrophoresis in polyacrylamide gels containing SDS, the radioactivity that is incorporated in a Ca^{2+}-dependent fashion appears in a single band (Figure 4) corresponding to a polypeptide with M_r of about 140,000.[7] As mentioned above, studies with purified Ca^{2+}-ATPase preparations have established that this polypeptide is the ATPase itself.

The stability of the denatured phosphoenzyme is maximal near pH 1, and progres-

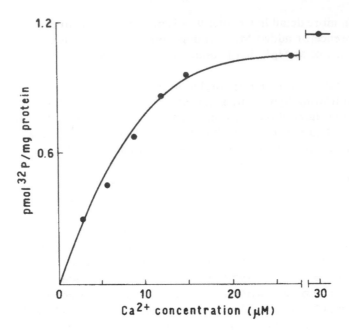

FIGURE 2.   The relation between the steady-state level of phosphorylation of human red blood cell membranes and the concentration of
Ca^{2+} at 0°C. (From Rega, A. F. and Garrahan, P. J., *J. Membr. Biol.*,
22, 313, 1975. With permission.)

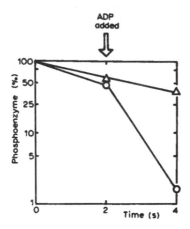

FIGURE 3.   The effects of ADP in the presence (O) and
absence (Δ) of Ca^{2+} on the phosphoenzyme from human red
blood cell membranes in the absence of Mg^{2+}. (From Rega,
A. F. and Garrahan, P. J., *Biochim. Biophys. Acta*, 507,
182, 1978. With permission.)

sively decreases as pH raises (Figure 5).[9,24] Hydroxylamine[9,24] and molybdate[24] accelerate the rate of cleavage of phosphate from the denatured phosphoenzyme. These
properties suggest that phosphate is associated to the enzyme through an acylphosphate bond.

In the phosphoenzyme of the Ca^{2+}-ATPase of sarcoplasmic reticulum[25] and of the
(Na^+, K^+)-ATPase,[1] the acylphosphate bond has been identified with aspartylphosphate. No chemical identification of the bond has yet been performed in the plasma

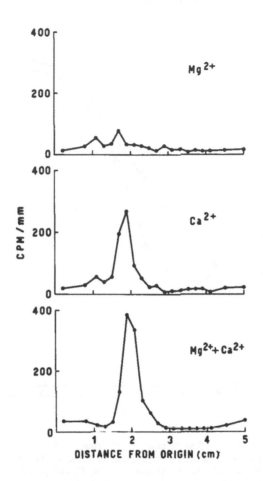

FIGURE 4.   Distribution of radioactivity in polyacrylamide gels after electrophoresis of erythrocyte membranes
phosphorylated with (^{32}P-γ)ATP in media containing Mg^{2+},
Ca^{2+}, or Mg^{2+} plus Ca^{2+}. (From Rega, A. F. and Garrahan,
P. J., *J. Membr. Biol.*, 22, 313, 1975. With permission.)

membrane Ca^{2+}-ATPase. However, we have shown that the limit phosphopeptides obtained above digestion of phosphorylated red blood cell membranes with pronase show
the same electrophoretic mobility as those of the Ca^{2+}-ATPase of sarcoplasmic reticulum and of the (Na$^+$, K$^+$)-ATPase of plasma membranes.[26] This suggests that the amino
acid sequence around the active site of the three ATPases is the same, a fact which is
in keeping with the close structural and functional homologies of all cation transport-
ATPases.

C. Dephosphorylation

The phosphoenzyme formed during Reaction 1 is able to donate its phosphate to
water according to the reaction:

$$E{\sim}P + H_2O \rightarrow E + Pi \qquad (2)$$

Reaction 2 can be measured following the time-course of the decay of the ^{32}P-labeled
phosphoenzyme after phosphorylation is stopped (Figure 6). To avoid the contribution
of the reversal of Reaction 1 to the observed rate of dephosphorylation it is advisable
to perform the measurements in media with either low concentrations of ADP or without Ca^{2+}.

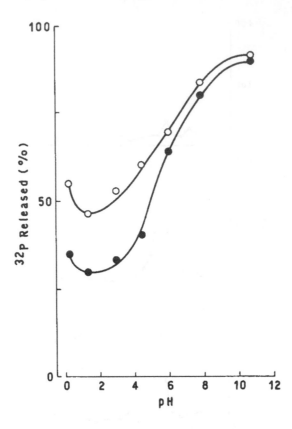

FIGURE 5.   The release of ³²P from TCA-denatured phosphoenzyme after incubation at 40°C during 30 min in media of different pH. (From Rega, A. F. and Garrahan, P. J., *J. Membr. Biol.*, 22, 313, 1975. With permission.)

Reaction 2 has been studied in considerable detail.[9,14] The following are its salient features:

1.   Dephosphorylation is enzymatically catalyzed, since the half-time for phosphoenzyme hydrolysis at 0°C increases from a few seconds to 10 to 20 min after the membrane proteins are denatured.[9]

2.   The rate of hydrolysis is not affected by Ca^{2+} in the 0 to 50 μM concentration range.[9] Ca^{2+} at concentrations in the millimolar range inhibits dephosphorylation,[17] as it inhibits the Ca^{1+}-ATPase activity (see Chapter 6, Section II).

3.   Neither ATP nor Mg^{2+} alone have any effect on the rate of phosphoenzyme hydrolysis, but when ATP and Mg^{2+} are added together, the rate of the reaction increases 5 to 10 times (see References 9 and 14 and Figure 7).

4.   Stimulation of phosphoenzyme hydrolysis in the presence of Mg^{2+} requires ATP at concentrations much higher than those needed for phosphorylation or for full occupation of the high-affinity site of the ATPase.[14]

5.   If present during phosphorylation, Mg^{2+} can be removed by chelators at the beginning of dephosphorylation without altering the stimulatory effect of ATP on the hydrolysis of the phosphoenzyme (Figure 8 and Reference 14). Hence, Mg^{2+} and ATP seem to act at different steps of the reaction.

6. La^{3+}, which does not affect phosphorylation, blocks the effect of ATP + Mg^{2+} on dephosphorylation. Inhibition is lost if La^{3+} is added after the enzyme has been phosphorylated in media with Mg^{2+}.[19,27]

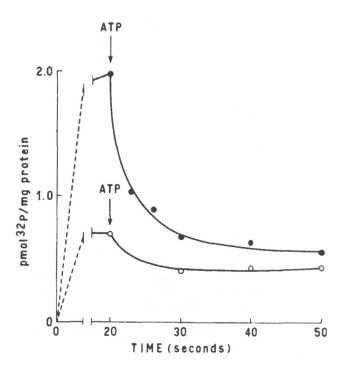

FIGURE 6. Time course of the release of ³²P from erythrocyte membranes phosphorylated with (³²P-γ)ATP in media with (●) and without (O) Ca²⁺ at 0°C. The incorporation of ³²P was stopped by the addition of an excess of unlabeled ATP. (From Rega, A. F. and Garrahan, P. J., *J. Membr. Biol.*, 22, 313, 1975. With permission.)

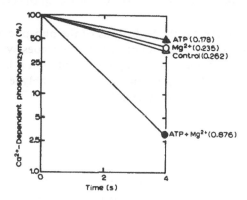

FIGURE 7. Dephosphorylation of the phosphoenzyme made in the absence of MgCl₂ in control medium and in medium containing 1 m*M* ATP, 0.5 m*M* MgCl₂ or 1 m*M* ATP plus 1.5 m*M* MgCl₂. The figures in parenthesis are rate constants for dephosphorylation in sec⁻¹. (From Garrahan, P. J. and Rega, A. F., *Biochim. Biophys. Acta*, 513, 59, 1978. With permission.)

1. The $E_1 \smile P \rightleftharpoons E_2 \smile P$ Transition

We have proposed[9,14,26] that most of the effects of Mg²⁺ and of ATP on dephosphorylation can be explained if we assume that Reaction 2 involves two successive elementary steps: the first is a conformational transition that drives the phosphoen-

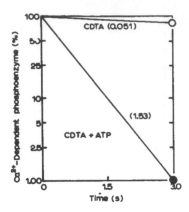

FIGURE 8.   The effect of removal of Mg²⁺ with 20 mM CDTA on dephosphorylation of human red blood cell membranes in media with and without 1 mM ATP. Figures in parenthesis are rate constants for dephosphorylation. (From Garrahan, P. J. and Rega, A. F., *Biochim. Biophys. Acta*, 513, 59, 1978. With permission.)

zyme from a state $E_1 \smallfrown P$ without altering its covalent structure. This step would be promoted by Mg²⁺ and would not require ATP.

$$E_1 \sim P \; \overset{Mg^{2+}}{\rightleftharpoons} \; E_2 \sim P \tag{3}$$

The terms $E_1 \smallfrown P$ and $E_2 \smallfrown P$ used to design the two conformers of the phosphoenzyme are based on the nomenclature made current by studies on the elementary steps of the (Na⁺, K⁺)-ATPase.[1]

The second step of Reaction 2 is the hydrolysis of the phosphoenzyme. Our assumption regarding this step is that $E_2 \smallfrown P$, but not $E_1 \smallfrown P$, undergoes hydrolysis and that once Mg²⁺ has acted on Reaction 3, the rate of dephosphorylation is largely increased when ATP is bound at a site in $E_2 \smallfrown P$ whose affinity is considerably less than that of the site at which ATP binds to promote phosphorylation, i.e.,

$$E_2 \sim P + H_2O \; \overset{ATP}{\rightleftharpoons} \; E_2 + Pi \tag{4}$$

At low concentrations of ATP, its site in $E_2 \smallfrown P$ would be practically empty, Reaction 4 would proceed at a low rate, and Mg²⁺ will be ineffective. Since catalysis of ATP hydrolysis can be fully explained by the transfer of the γ-phosphate of the nucleotide to the enzyme in Reaction 2 followed by its transference to water in Reaction 4, ATP would act in $E_2 \smallfrown P$ as a modifier without being hydrolyzed.

The reaction scheme in Equations 3 and 4 provides an explanation for the mechanism of inhibition by La³⁺. In fact, if La³⁺ blocked the $E_1 \smallfrown P \rightarrow E_2 \smallfrown P$ transition, it is easy to see that, in agreement with the experimental results, La³⁺ will prevent acceleration of hydrolysis by Mg²⁺ and ATP if added before Mg²⁺, but it will be ineffective if added after Mg²⁺ has accelerated the transformation of $E_1 \smallfrown P$ into $E_2 \smallfrown P$.

It is important to notice that since our assumption implies that $E_2 \smallfrown P$ is also formed in the absence of Mg²⁺, it is necessary to explain why, under these conditions, ATP is unable to activate hydrolysis of the phosphoenzyme. Two mechanisms may account for this effect: (1) Mg⁺ could displace the steady-state distribution between conformers towards $E_2 \curvearrowright P$ and/or, (2) $E_2 \curvearrowright P$ formed in the presence of Mg^{2+} could have different reactivity than that formed in its absence, as if Mg^{2+} would have remained in $E_2 \curvearrowright P$ in a tightly bound or "occluded" state.

It is tempting to think that the effect of ATP at the low-affinity site in $E_2 \curvearrowright P$ is expressed in the overall reaction as the low-affinity component of the biphasic substrate curve of the Ca^{2+}-ATPase. Two experimental facts however, complicate this interpretation.[18] First: in the overall reaction the low affinity component of the substrate curve seems to be observable only in enzymes bound with calmodulin, whereas stimulation of $E_2 \curvearrowright P$ hydrolysis by high ATP concentrations is also present in calmodulin-stripped membranes.[18] Second: in the overall reaction, only the MgATP complex seems to be the effective ligand,[18] whereas in Reaction 4, free ATP is as effective as MgATP in promoting hydrolysis of $E_2 \curvearrowright P$. Further experimental data are therefore needed before deciding whether these discrepancies are a consequence of the different conditions under which ATPase activity and elementary steps are assayed or they are caused by other, and yet unidentified, processes.

D. The $E_2 \rightleftharpoons E_1$ Transition

Although the catalysis of ATP hydrolysis ends with Reaction 4, the enzymatic cycle is only completed when E_2 returns to E_1, i.e.,

$$E_2 \rightleftharpoons E_1 \tag{5}$$

Reaction 5 is the less known of the elementary steps of the Ca^{2+}-ATPase. Some light has been thrown on it by the experiments of Muallem and Karlish.[28] These authors studied the irreversible inhibition of the Ca^{2+}-ATPase of red blood cell membranes by fluorescein isothiocyanate (FITC) (see Chapter 11, Section IV) a reagent that binds covalently to the sites of the enzyme for ATP. The time course of inhibition by FITC can be described by the sum of two exponential terms, the apparent affinity for inhibition by FITC being higher for the faster term. ATP protects against FITC with two distinctly separate affinities ($K_{0.5}$ about 5 and 180 μM). These results were interpreted as indicating that the enzyme exists in at least two conformational states which are similar or identical to the E_1 and E_2 states that participate in ATP hydrolysis. E_1 would possess the high- and E_2 the low-affinity sites for ATP and FITC. Since the experiments of Muallem and Karlish were performed in the absence of Mg^{2+}, they provide evidence for the existence of E_2 in the absence of the cation which implies that E_2 is an independent entity and not just the product of the reaction of the enzyme with Mg^{2+}.

The kinetic analysis of inhibition indicates that the two sites for ATP do not coexist in the same enzyme molecule, but are different states of the same site. On this hypothesis, the existence of two different rates of inactivation and of two different affinities for ATP as a protector against inactivation necessarily requires that during treatment with FITC the rate of interconversion between E_1 and E_2 be slow, compared to the rate of combination with the inhibitor. This contrasts with the high rate of interconversion that has to be postulated if Reaction 5 is to be considered a part of the catalytic cycle of the ATPase. Muallem and Karlish[28] suggested that during enzymatic activity, Reaction 5 proceeds at a high rate, because this reaction is accelerated by ATP acting at the site in E_2 where it protects against inhibition by FITC. Although there is no direct experimental evidence in favor of this view, acceleration by ATP of the $E_2 \rightleftharpoons E_1$ transition has been demonstrated in the (Na^+, K^+)-ATPase[1] and in the Ca^{2+}-ATPase of sarcoplasmic reticulum.[5]

Table 1

EFFECTS OF Ca²⁺, Mg²⁺, AND ATP ON THE ELEMENTARY STEPS OF THE Ca²⁺-ATPase

Step	Ca²⁺	ATP	Mg²⁺	Rate
$E_1 + ATP \rightarrow E_1 P + ADP$	Essential	Essential with high affinity	Activates	Low at 0 mM Mg²⁺
$E_1 P \rightarrow E_2 P$	Not needed	Not needed	Activates	Low at 0 mM Mg²⁺
$E_2 P \rightarrow E_2 + Pi$	Not needed	Activates with low affinity	Essential in the previous step for activation by ATP	Low without ATP
$E_2 \rightarrow E_1$	Not needed	Activates with low affinity?	Activates?	Low without ATP?

II. REACTION SCHEME FOR THE HYDROLYSIS OF ATP

Table 1 summarizes the dependence on ligands of the elementary steps of the hydrolysis of ATP by the Ca²⁺-ATPase. Figure 9 shows a reaction scheme for the hydrolysis of ATP catalyzed by this enzyme. The scheme has been formulated combining our knowledge on the elementary steps and steady-state kinetics of the Ca²⁺-ATPase with what is known of the other cation-transport ATPases. The reaction scheme is based on seven main assumptions, namely:

1. The enzyme exists in two conformers: E_1 and E_2.
2. Hydrolysis of ATP proceeds through the phosphoenzymes $E_1 \sim P$ and $E_2 \sim P$.
3. E_1 possesses a high-affinity catalytic site for ATP. In E_2 this site is converted into a low-affinity noncatalytic site.
4. E_1 catalyzes the transphosphorylation between the enzyme and ADP, and E_2 catalyzes the transphosphorylation between the enzyme and water.
5. The enzymatic activity of E_1 is expressed only in the presence of Ca²⁺.
6. The catalytic activity of E_2 does not require Ca²⁺ and, if Mg²⁺ has reacted with the enzyme, is strongly activated by ATP at the low affinity site.
7. ATP at the low affinity site may also accelerate the E_2 to E_1 transition.

These properties predict quantitatively the overall steady-state behavior of the Ca²⁺-ATPase activity since

1. In the absence of added Mg²⁺, only the high affinity site in E_1 would be operative. In agreement with this, under these conditions, ATPase activity is low and the substrate curve of the Ca²⁺-ATPase seems to be a single Michaelis equation of low maximum velocity (less than 5% of the optimal) and high apparent affinity (Km $\sim$ 10 μM).
2. In the presence of Mg²⁺ and enough ATP to fully activate phosphorylation, very little activation by ATP at E_2 is to be expected. In agreement with this, under these conditions, ATPase activity, although higher than in the absence of Mg²⁺, is still less than 25% of the optimal activity.

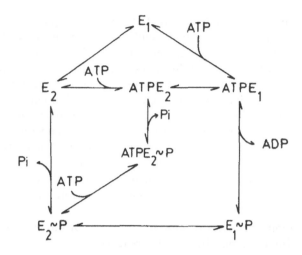

FIGURE 9. Reaction scheme for the hydrolysis of ATP by the Ca^{2+}-ATPase of the plasma membrane. The requirements of ligands other than ATP of each of the elementary steps are indicated in Table 1. Both at high and low concentrations of ATP the reaction pathway is the same up to $E_2 \smallfrown P$. At low concentrations of ATP the reaction proceeds through the pathway:

$$E_2 \smallfrown P \rightarrow E_2 \rightarrow E_1 \rightarrow E_1 ATP$$

As the concentration of ATP raises, this pathway is progressively replaced by the following:

$$E_2 \smallfrown P \rightarrow ATP\ E_2 \smallfrown P \rightarrow ATPE_2 \rightarrow ATPE_1$$

which is faster because ATP accelerates dephosphorylation and perhaps also because the $ATPE_2 \rightarrow ATPE_1$ transition is faster than the $E_2 \rightarrow E_1$ transition. At nonlimiting concentrations of ATP the whole metabolic flux takes place through this pathway.

3. Full activation in the presence of Mg^{2+} requires ATP at high concentrations to activate $E_2 \smallfrown P$ hydrolysis and perhaps also the E_2 to E_1 transition. In agreement with that, ATPase activity in the presence of Mg^{2+} is high and the response to ATP is biphasic with a high-affinity low-velocity and a low-affinity and high-velocity component (see Chapter 6, Section III).

III. ENERGY CHANGES DURING THE ELEMENTARY STEPS

Very little experimental work has been done on this important problem. Schatzmann and Luterbacher[29] approached the subject measuring phosphorylation in red blood cell membranes and in purified ATPase at 0°C in media with La^{2+}. Since La^{2+} blocks the $E_1 \smallfrown P \rightleftharpoons E_2 \smallfrown P$ transition, phosphorylation could be studied as an equilibrium of the form:

$$E_1 \; + \; ATP \overset{k_1}{\underset{k_2}{\rightleftharpoons}} E_1 {\sim} P \; + \; ADP \tag{6}$$

Reaction 6 described the initial and final states of a sequence of reactions which involve at least the binding of ATP to E_1 and the transphosphorylation reaction. Its equilibrium constant therefore will be the product of at least two equilibrium constants and the Gibbs energy changes will be the sum of at least two individual changes.

If Equation 6 described the observed behavior of the Ca²⁺-ATPase, the ratio of the rate constants k_1 and k_2 estimated from the phosphorylation curve that gives the best fit to the experimental results could be used to calculate the equilibrium constant of the reaction. On the basis of this assumption, Schatzmann and Luterbacher estimated an equilibrium constant of 1.8×10^{-2} and 2×10^{-3} for isolated membranes and purified enzyme, respectively, corresponding to Gibbs energy changes of 9.1 kJ/mol and 14.0 kJ/mol. This suggests that phosphorylation is strongly endergonic, a fact that is in agreement with our observation that the phosphoenzyme is dephosphorylated by ADP.[23]

For the case of dephosphorylation, the overall equilibrium constant and Gibbs energy change was estimated from the two step reaction:

$$E_1 {\sim} P \rightleftharpoons E_2 {\sim} P \rightleftharpoons E_2 + Pi \qquad (7)$$

Gibbs energy change was calculated subtracting from the Gibbs energy change for ATP hydrolysis the value estimated for Reaction 6. This gives about −40 kJ/mol for the energy change and about 10^8 M for the equilibrium constant, both in isolated membranes and purified enzyme. Therefore the steps involved in dephosphorylation seem to be strongly exergonic.

IV. THE PHOSPHATASE ACTIVITY OF THE Ca²⁺-ATPase

The Ca²⁺-ATPase of plasma membranes, like the Ca²⁺-ATPase of sarcoplasmic reticulum[30] and the (Na⁺ + K⁺)-ATPase[31], shows a phosphatase activity towards *p*-nitrophenylphosphate. It is not yet clear if the phosphatase activity of transport ATPases has a physiological role or if it only expresses the ability of these enzymes to catalyze the hydrolysis of artificial substrates in a physiologically meaningless reaction. Regardless of the uncertainties about its biological significance, studies on this activity have been useful in the characterization of the elementary steps of the ATPase reaction.

A. General Properties

This activity was discovered in our laboratory when we found, during studies of the K⁺-dependent phosphatase of the Na⁺ pump, that the overall K⁺-phosphatase activity of red blood cell membranes was strongly activated by ATP and an endogenous factor that could be removed by EGTA and that proved to be contaminating Ca²⁺.[32]

All studies on this activity have been performed in red blood cell membranes. No information about its presence in other plasma membranes is yet available.

The phosphatase activity of the Ca²⁺-pump requires Mg²⁺ and Ca²⁺, is strongly activated by K⁺ and Na⁺, and is absolutely dependent on the presence of ATP.[32-34] The dependence on ATP seems to be a unique feature of this phosphatase, since ATP is not needed for the phosphatase activities of the Na⁺ pump and of the sarcoplasmic reticulum Ca²⁺-ATPase.

B. Kinetics

1. The Substrate Curve

Only *p*-nitrophenylphosphate (pNPP) has been used as substrate. Monophosphate esters like acetylphosphate and carbamylphosphate, which are substrates for the phosphatase activity of other transport ATPases,[31] have not been tested in the Ca²⁺-ATPase of plasma membranes.

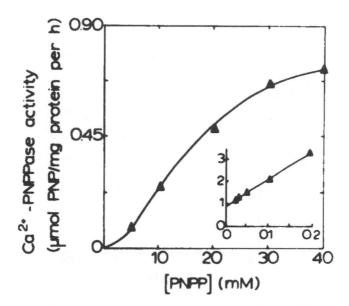

FIGURE 10.   Ca²⁺-p-nitrophenylphosphatase activity of red blood cell membranes as a function of pNPP concentration. The inset is a plot of the reciprocal of the square root of Ca²⁺-pNPPase activity as a function of the reciprocal of pNPP concentration. (From Caride, A. J., Rega, A. F., and Garrahan, P. J., *Biochim. Biophys. Acta*, 689, 421, 1982. With permission.)

At low ATP concentrations (30 μM), the activity first increases and then decreases tending to zero as the concentration of pNPP rises.[35] Inhibition by excess pNPP disappears as the concentration of ATP increases, suggesting that its cause is the displacement by pNPP of ATP from the site at which the nucleotide promotes phosphatase activity. This point will be discussed in more detail below.

In media with high ATP and optimal concentrations of cofactors, phosphatase activity increases with the concentration of pNPP along an S-shaped curve (Figure 10), which can be adjusted to the equation:

$$v = \frac{Vm}{(1 + Kp/[pNPP])^2} \tag{8}$$

At 8 mM ATP, Kp is about 13 mM. The low apparent affinity for pNPP makes it impractical to assay the activity at saturating pNPP concentrations. For this reason, experimentally measured phosphatase activities are much lower than ATPase activities. However, the extrapolated maximum rate (Vm in Equation 8) is very close to the maximum rate of the ATPase reaction, indicating that the turnover of the Ca²⁺-ATPase during ATPase and phosphatase activities is about the same. If we assume equilibrium kinetics, Equation 8 would indicate that to elicit phosphatase activity, pNPP must occupy two identical and noninteracting sites in each enzyme molecule. However, in view of the complex multistage reaction mechanism of the Ca²⁺-ATPase, other nonequilibrium mechanisms may provide equally plausible interpretations of the shape of the curve.

2. Dependence on Mg²⁺

The phosphatase activity of the Ca²⁺ pump is absolutely dependent on Mg²⁺. This property is shared with the phosphatases of other transport ATPases. The value of the

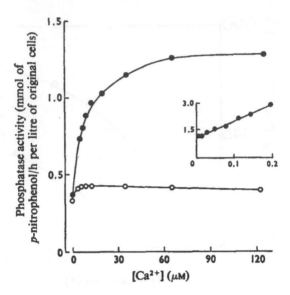

FIGURE 11.   Effect of Ca²⁺ on *p*-nitrophenylphosphatase of red blood cell membranes in the presence (●) and absence (○) of 100 mM KCl in media with 5 mM MgCl₂. The inset is a double reciprocal plot of Ca²⁺-dependent phosphatase activity in the presence of KCl vs. Ca²⁺ concentration. (From Rega, A. F., Richards, D. E., and Garrahan, P. J., *Biochem. J.*, 136, 185, 1973. With permission.)

K_m for activation of the phosphatase (about 0.5 to 1 mM) is higher than that for activation of the ATPase and close to the K_i for the low-affinity inhibition of the ATPase by Mg²⁺.[36] A similar effect of Mg²⁺ has been reported for the phosphatase of the Na⁺ pump of red blood cell membranes.[37]

3. Activation by Ca²⁺

Ca²⁺ is essential for phosphatase activity (Figure 11). The value of K_{Ca} for activation of the phosphatase is almost identical to that for activation of the ATPase.[34] Sr²⁺ and Ba²⁺ replace Ca²⁺, and the stimulation by these cations takes place along curves that are superimposable with those for stimulation of the ATPase.[33] Studies on reconstituted ghosts of red blood cells show that phosphatase requires, like ATPase activity, Ca²⁺ at the inner surface of the cell membrane.[38]

The features in common shared by the kinetics of activation of phosphatase and of ATPase activities by Ca²⁺ strongly suggest that the site from which Ca²⁺ promotes phosphatase activity is the site for Ca²⁺ transport of the Ca²⁺-ATPase.

4. Effects of Monovalent Cations

The phosphatase activity of the Ca²⁺ pump is strongly stimulated by K⁺ (Figure 11) and by Na⁺. The activity in the absence of these cations is less than 5% of the optimal. Na⁺ and K⁺ have the same maximum effect, but the apparent affinity for activation by K⁺ (35 mM) is much higher than that for activation by Na⁺ (190 mM).[34]

The large effect of K⁺ on the phosphatase activity of the Ca²⁺-ATPase of red blood cells makes it mandatory to cancel from the measured activity the contribution of the activity of the K⁺-dependent phosphatase of the Na⁺ pump also present in these membranes. This activity does not require Ca²⁺ or ATP and is inhibited by cardiac glycosides.[39]

Monovalent cations also stimulate the overall Ca²⁺-ATPase activity (see Chapter 9);

the order of apparent affinities for this effect is similar to that of the phosphatase. Activation of the phosphatase, however, differs in two respects from the activation of the ATPase. First: the overall reaction is considerably less dependent on monovalent cations than the phosphatase, maximum activation of the Ca^{2+}-ATPase activity by a monovalent cation being about 30 to 40%; and second: the apparent affinities for activitation of the ATPase are about fivefold higher than those for activation of the phosphatase.

C. The Interaction Between the Sites for pNPP and the Sites for ATP

We have already mentioned that the strict dependence on ATP is a salient and perhaps unique property of the phosphatase activity of the Ca^{2+}-ATPase of the plasma membrane.

The effect is highly specific for ATP. CTP, GTP, ITP, or UTP neither replace nor interfere with ATP.[34] ATP is only effective at the inner surface of the cell membrane.[38] Since the sites for ATP as promoter of phosphatase activity show the same sideness and specificity than the sites for ATP of the Ca^{2+}-ATPase, it is reasonable to think that ATP promotes phosphatase activity from the same sites that participate in ATPase activity.

ATP activates phosphatase activity along a single Michaelis equation. This contrasts with the biphasic response to ATP of the ATPase and suggests that only one of the two classes of site for ATP of the ATPase is involved in the promotion of phosphatase activity.

1. The High-Affinity Site

The apparent affinity of the site from which ATP promotes phosphatase activity is not significantly different from that of the high-affinity site of the ATPase. In both, pNPP acts as a competitive inhibitor with the same Ki value (Figure 12). These results strongly suggest that the site from which ATP promotes phosphatase activity is the high-affinity site of the ATPase.[35]

2. The Low-Affinity Site

Low-affinity activation of the ATPase by ATP and ATP-dependent phosphatase activity seem to be mutually exclusive functions of the Ca^{2+} pump. This implies that those ATPase units that catalyze pNPP hydrolysis will not exhibit low-affinity activation by ATP, and conversely, those units in which ATP stimulates ATP hydrolysis will not catalyze pNPP hydrolysis.

This assertion is supported by the observation that low-affinity stimulation of ATPase by ATP is associated with a parallel inhibition of phosphatase activity, whereas stimulation of phosphatase by pNPP is accompanied by a parallel inhibition of the low-affinity activation of the ATPase by ATP (Figure 13A and B).[35]

The interactions between pNPP and ATP at the low-affinity site can be accounted for if we assume that the low-affinity site for ATP is the active site for pNPP hydrolysis, pNPP being unable to substitute for ATP as activator of the ATPase.[35]

3. ATP Hydrolysis During Phosphatase Activity

The requirement for ATP at the high-affinity site for phosphatase activity raises the question of whether or not ATP hydrolysis at the high-affinity site proceeds concurrently with catalysis of pNPP hydrolysis at the low-affinity site. Experimental evidence seems to provide an affirmative answer to this question. In fact, if ATPase activity is measured as a function of the concentration of pNPP, and the levels of ATP and of pNPP are adjusted in such a way as to keep the high-affinity site saturated with ATP, the activity of the ATPase will not tend to zero, but to a value close to the activity

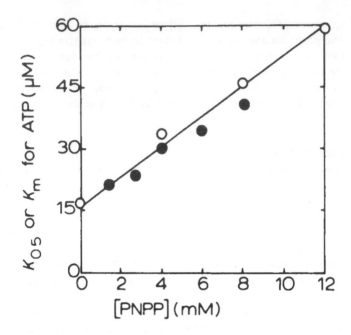

FIGURE 12.   The effect of pNPP of the $K_{0.5}$ for ATP as activator of the Ca²⁺-pNPP (●) and the Km of the high-affinity site of the Ca²⁺-ATPase (O) from human red blood cell membranes. (From Caride, A. J., Rega, A. F., and Garrahan, P. J., *Biochim. Biophys. Acta,* 689, 421, 1982. With permission.)

attainable when only the high-affinity site participates in ATP hydrolysis (Figure 13B). Since under these conditions phosphatase activity is approaching its maximum value, it seems difficult to avoid the conclusion that phosphatase and ATPase activities are proceeding simultaneously.[35]

Table 2 summarizes our present ideas on the interactions between pNPP and ATP. Eight different complexes between the ATPase, pNPP, and ATP would be possible if the high- and low-affinity sites for ATP of the Ca²⁺-ATPase were also able to bind pNPP.

We have already discussed the catalytic properties of the complexes of the enzyme with ATP alone and provided evidence that the high-affinity site is the catalytic site for ATP. The results discussed here and summarized in Table 2 suggest that phosphatase activity is only possible if ATP is bound to this site. This would explain why none of the complexes with pNPP alone is catalytically effective. An intriguing question raised by our scheme is why the low-affinity site, which is usually considered to be a noncatalytic effector site for ATP, becomes a catalytic site when pNPP binds to it. In this respect, it is interesting to point out that the low-affinity, noncatalytic site for ATP of the (Na⁺, K⁺)-ATPase has also been proposed as the active site for the phosphatase activity of this enzyme.[40]

In Table 2 we have assumed that net turnover of the ATPase is taking place during phosphatase activity. Although there is experimental evidence for this view, it is not yet clear if this is required for phosphatase activity. If net turnover were, in fact, required, it becomes necessary to explain why ATP-dependent phosphatase activity persists after most of the Ca²⁺-ATPase and the Ca²⁺-dependent phosphorylation have been abolished by treatment of red blood cell membranes with phospholipase C.[41] If the only effect of this treatment were to inactivate phosphorylation, this result would imply that binding of ATP without any subsequent chemical reaction is all that is

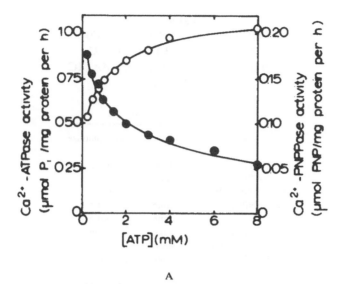

A

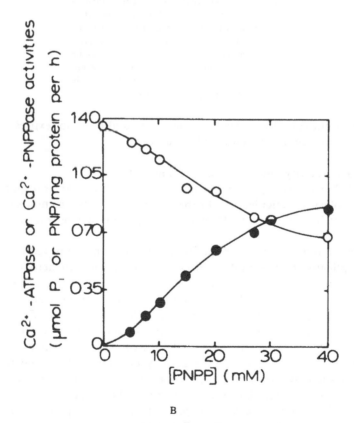

B

FIGURE 13.   Ca²⁺-ATPase (O) and Ca²⁺-*p*-nitrophenylphosphatase
(●) activities as a function of ATP (a) and pNPP (b) concentrations.
(From Caride, A. J., Rega, A. F., and Garrahan, P. J., *Biochim.
Biophys. Acta*, 689, 421, 1982. With permission.)

needed for the E_2 conformer to acquire catalytic activity towards pNPP. However, in
view of the strong effect that the phospholipid environment has on the function of the
Ca²⁺-ATPase (see Chapter 10), we cannot discard the possibility that this response is

Table 2

THE EIGHT COMPLEXES OF
ATP AND pNPP WITH THE Ca^{2+}
ATPase AND THEIR ENZYMATIC
ACTIVITIES

	Activity	
Complex	ATPase	Phosphatase
(1) EATP$_h$	slow	none
(2) EATP$_l$	none	none
(3) EATP$_h$ATP$_l$	fast	none
(4) EpNPP$_h$	none	none
(5) EpNPP$_l$	none	none
(6) EpNPP$_h$pNPP$_l$	none	none
(7) EATP$_h$pNPP$_l$	slow	fast
(8) EpNPP$_h$ATP$_l$	none	none

Note: The subscripts (h) and (l) correspond to
ligands bound to the high and low affinity sites for ATP, respectively. Since the
sites for ATP seem to be different states
of the same site, the ligands would not be
simultaneously bound to the same enzyme unit, but would appear sequentially
bound during a hydrolysis cycle.

peculiar only to phospholipase-treated membranes and does not express the behavior
of the enzyme under more physiological conditions.

D. Phosphatase Activity and Active Ca^{2+} Transport

It is known that pNPP hydrolysis is able to drive active Ca^{2+} transport in the Ca^{2+}
pump of sarcoplasmic reticulum.[30] On the other hand, experimental evidence[42] indicates that pNPP hydrolysis is ineffective in supporting active transport by the Na$^+$
pump.

The experimental analysis of this question in the Ca^{2+} pump of red blood cell membranes is complicated by the ATP requirements of phosphatase activity. For this reason, the detection of pNPP-dependent Ca^{2+} pumping requires a comparison of the
effects of pNPP on ATP hydrolysis and on active Ca^{2+} transport under conditions in
which pNPP does not displace ATP from the high affinity site. From this sort of
experiment, three kinds of results can be predicted: (1) if pNPP hydrolysis promoted
Ca^{2+} transport with the same efficiency as ATP hydrolysis, increments in the concentration of pNPP should inhibit ATPase activity but not Ca^{2+} transport; (2) if pNPP
hydrolysis promoted transport, but with less effectiveness than ATP hydrolysis, Ca^{2+}
transport would be inhibited, but to a lesser degree than ATPase activity; and finally,
(3) if pNPP hydrolysis were unable to promote transport, pNPP should inhibit in
parallel Ca^{2+}-ATPase activity and active Ca^{2+} transport.

We have performed experiments based on this reasoning using resealed ghosts[34] and
inside-out vesicles of red blood cell membranes[43] (Figure 14). In both preparations as
a function of pNPP concentration, Ca^{2+}-ATPase activity and Ca^{2+} pumping are inhibited along curves that are not significantly different. This constitutes strong evidence
that hydrolysis of pNPP is unable to energize Ca^{2+} transport.

These results seem to indicate that hydrolysis of pNPP by the Ca^{2+} pump takes place
using a part of the reaction pathway of ATP hydrolysis that is insufficient to energize

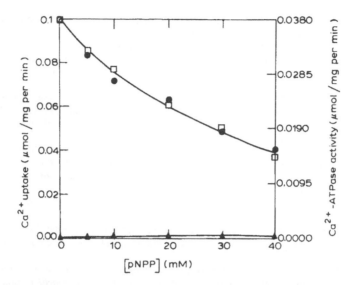

FIGURE 14.   Ca uptake in the presence (●) and absence (▲) of ATP
and Ca²⁺-ATPase activity (□) in IOVs from human red blood cells as
a function of pNPP concentration. (From Caride, A. J., Rega, A. F.,
and Garrahan, P. J., *Biochim. Biophys. Acta*, 734, 363, 1983.)

ion translocation. The ineffectiveness of pNPP suggests that the mechanism of phosphatase activity in plasma membrane Ca²⁺-ATPase may be different from that of sarcoplasmic reticulum and closer to that of the plasma membrane (Na⁺, K⁺)ATPase.

REFERENCES

1. Schuurmans, F. M. A. H. and Bonting, S. L., Sodium-potassium activated adenosine triphosphatase, in *Membrane Transport,* Bonting, S. L. and de Pont, J. J. H. H. M., Eds., Elsevier/North-Holland, Amsterdam, 1981, chap. 6.
2. Hasselbach, W., Calcium-activated ATPase of the sarcoplasmic reticulum, in *Membrane Transport,* Bonting, S. L. and de Pont, J. J. H. H. M., Eds., Elsevier/North-Holland, Amsterdam, 1981, chap. 8.
3. de Pont, J. J. H. H. M. and Bonting, S. L., Anion sensitive ATPase and (K⁺ + H⁺)-ATPase, in *Membrane Transport,* Bonting, S. L. and de Pont, J. J. H. H. M., Eds., Elsevier/North-Holland, Amsterdam, 1981, chap. 8.
4. Mitchell, P., Chemiosmotic ATPase mechanisms, *Ann. N.Y. Acad. Sci.,* 402, 584, 1982.
5. de Meis, L., The sarcoplasmic reticulum, transport and energy transduction, in *Transport in the Life Sciences,* Vol. 2, Bittar, E. E., Ed., John Wiley & Sons, New York, 1981, chap. 8.
6. Froelich, J. P., Hobbs, A. S., and Albers, W., Evidence for parallel pathways of phosphoenzyme formation in the mechanism of ATP hydrolysis by electrophorus Na,K-ATPase, *Curr. Top. Membrane Transport,* 19, 513, 1983.
7. Knauf, P. A., Proverbio, F., and Hoffman, J. E., Electrophoretic separation of different phospho-proteins associated with Ca-ATPase and Na,K-ATPase in human red cell ghosts, *J. Gen. Physiol.,* 69, 324, 1974.
8. Katz, S. and Blonstein, R., Ca²⁺-stimulated membrane phosphorylation and ATPase activity of the human erythrocyte, *Biochim. Biophys. Acta,* 389, 314, 1975.
9. Rega, A. F. and Garrahan, P. J., Calcium ion-dependent phosphorylation of human erythrocyte membranes, *J. Membr. Biol.,* 22, 313, 1975.
10. Niggli, V., Penniston, J. T., and Carafoli, E., Purification of the (Ca²⁺ - Mg²⁺)-ATPase from human erythrocyte membranes using a calmodulin-affinity column, *J. Biol. Chem.,* 254, 9955, 1979.
11. Lichtner, R. and Wolf, H. U., Characterization of the phosphorylated intermediate of the isolated high-affinity (Ca²⁺ + Mg²⁺)-ATPase of human erythrocyte membranes, *Biochim. Biophys. Acta,* 598, 486, 1980.

12. Stieger, J. and Luterbacher, S., Some properties of the purified (Ca²⁺ + Mg²⁺)-ATPase from human red cell membranes, *Biochim. Biophys. Acta,* 641, 270, 1981.
13. Szász, I., Hasitz, M., Sarkadi, B., and Gárdos, G., Phosphorylation of the Ca²⁺ pump intermediate in intact red cells, isolated membranes and inside-out vesicles, *Mol. Cell Biochem.,* 22, 147, 1978.
14. Garrahan, P. J. and Rega, A. F., Activation of the partial reactions of the Ca²⁺-ATPase from human red cells by Mg²⁺ by ATP, *Biochim. Biophys. Acta,* 513, 59, 1978.
15. Richards, D. E., Rega, A. F., and Garrahan, P. J., ATPase and phosphatase activities from human red cell membranes. I. The effects of N-ethylmaleimide, *J. Membr. Biol.,* 35, 113, 1977.
16. Schatzmann, H. J., Role of magnesium in the (Ca²⁺ + Mg²⁺)-stimulated membrane ATPase of human red cells, *J. Membr. Biol.,* 35, 149, 1977.
17. Lichtner, R. and Wolf, H. U., Phosphorylation of the isolated high-affinity (Ca²⁺ + Mg²⁺)-ATPase of the human erythrocyte membrane, *Biochim. Biophys. Acta,* 598, 472, 1980.
18. Muallem, S. and Karlish, S. J. D., Studies on the mechanism of regulation of the red cell Ca²⁺ pump by calmodulin and ATP, *Biochim. Biophys. Acta,* 647, 73, 1981.
19. Schatzmann, H. J. and Burgin, H., Calcium in human red blood cells, *Ann. N.Y. Acad. Sci.,* 307, 125, 1978.
20. de Smedt, H., Parys, J. B., Borghgraef, R., and Wuytack, C., Phosphorylated intermediates of (Ca²⁺ + Mg²⁺)-ATPase and alkaline phosphatase in renal plasma membranes, *Biochim. Biophys. Acta,* 728, 409, 1983.
21. Larocca, J. N., Rega, A. F., and Garrahan, P. J., Phosphorylation and dephosphorylation of the Ca²⁺ pump of human red cells in the presence of monovalent cations, *Biochim. Biophys. Acta,* 645, 10, 1981.
22. Schatzmann, H. J., The red cell calcium pump, *Ann. Rev. Physiol.,* 45, 303, 1983.
23. Rega, A. F. and Garrahan, P. J., Calcium ion-dependent dephosphorylation of the Ca²⁺-ATPase of red cells by ADP, *Biochim. Biophys. Acta,* 507, 182, 1978.
24. Lichtner, R. and Wolf, H. U., Characterization of the phosphorylated intermediate of the isolated high-affinity (Ca²⁺ + Mg²⁺)-ATPase of the human erythrocyte membrane, *Biochim. Biophys. Acta,* 598, 486, 1980.
25. Degani, C. and Boyer, P. D., A borohydride reduction method for the characterization of acyl phosphate linkage in proteins and its application to sarcoplasmic reticulum adenosine triphosphatase, *J. Biol. Chem.,* 248, 8222, 1973.
26. Rega, A. F., Garrahan, P. J., Barrabin, H., Horenstein, A., and Rossi, J. P., Reaction scheme for the Ca-ATPase from human red blood cells, in *Cation Flux across Biomembranes,* Mukohata, Y. and Packer, L., Eds., Academic Press, New York, 1979, 87.
27. Luterbacher, S. and Schatzmann, H. J., The site of action of La³⁺ in the reaction cycle of the human red cell membrane Ca²⁺ pump ATPase, *Experientia,* 39, 311, 1983.
28. Muallem, S. and Karlish, S. J. D., Catalytic and regulatory ATP-binding sites of the red cell Ca²⁺ pump studied by irreversible modification with fluorescein isothiocyanate, *J. Biol. Chem.,* 258, 169, 1983.
29. Luterbacher, S. and Schatzmann, H. J., Quantitative analysis of some of the partial reactions of the human red cell Ca²⁺-transport ATPase, *Experientia,* 39, 311, 1983.
30. Inesi, G., p-Nitrophenyl phosphate hydrolysis and calcium ion transport in fragmented sarcoplasmic reticulum, *Science,* 171, 901, 1971.
31. Rega, A. F. and Garrahan, P. J., Potassium activated phosphatase, in *The Enzymes of Biological Membranes. Membrane Transport,* Martonosi, A., Ed., Plenum Press, New York, 1976, chap. 12.
32. Pouchan, M. I., Garrahan, P. J., and Rega, A. F., Effects of ATP and Ca⁺⁺ on a K⁺-activated phosphatase from red cell membranes, *Biochim. Biophys. Acta,* 173, 151, 1969.
33. Garrahan, P. J., Pouchan, M. I., and Rega, A. F., Potassium activated phosphatase from human red cells. The effects of adenosine triphosphate, *J. Membr. Biol.,* 3, 26, 1970.
34. Rega, A. F., Richards, D. E., and Garrahan, P. J., Calcium ion-dependent p-nitrophenyl phosphate phosphatase activity and calcium ion dependent adenosine triphosphatase activity from human erythrocyte membranes, *Biochem. J.,* 136, 185, 1973.
35. Caride, A. J., Rega, A. F., and Garrahan, P. J., The role of the sites for ATP of the Ca²⁺-ATPase from human red blood cells during Ca²⁺-phosphatase activity, *Biochim. Biophys. Acta,* 689, 421, 1982.
36. Caride, A. J., Rega, A. F., and Garrahan, P. J., unpublished results.
37. Flatman, P. W. and Lew, V. L., Excess magnesium converts red cell (sodium + potassium) ATPase into the potassium phosphatase, *J. Physiol.,* 307, 1, 1980.
38. Rega, A. F., Garrahan, P. J., and Wainer, S. R., Asymmetrical activation by Ca²⁺ of the erythrocyte membrane K⁺-dependent phosphatase, *Experientia,* 28, 1158, 1972.
39. Garrahan, P. J., Pouchan, M. I., and Rega, A. F., Potassium activated phosphatase of human red cells. The mechanism of potassium activation, *J. Physiol.,* 202, 305, 1969.

40. Robinson, J. D., Levine, G. M., and Robinson, L. J., A model for the reaction pathway of the K$^+$-dependent phosphatase activity of the (Na$^+$ + K$^+$)-dependent ATPase, *Biochim. Biophys. Acta,* 731, 406, 1983.

41. Richards, D. E., Vidal, J. C., Garrahan, P. J., and Rega, A. F., ATPase and phosphatase activities from human red cell membranes. II. The effects of phospholipases on Ca^{2+}-dependent enzymic activities, *J. Membrane Biol.,* 35, 125, 1977.

42. Garrahan, P. J. and Rega, A. F., Potassium activated phosphatase from human red blood cells. The effects of *p*-nitrophenylphosphate on cation fluxes, *J. Physiol.,* 223, 595, 1972.

43. Caride, A. J., Rega, A. F., and Garrahan, P. J., Effects of *p*-nitrophenylphosphate on Ca^{2+} transport in inside-out vesicles from human red cell membranes, *Biochim. Biophys. Acta,* 734, 363, 1983.

Chapter 9

ACTIVATION BY MAGNESIUM AND BY ALKALI METAL IONS

P. J. Garrahan

I. MAGNESIUM

Like all other cation-transport ATPases, the Ca²⁺-ATPase of plasma membranes requires Mg²⁺ for optimal activity. All activating effects of Mg²⁺ are exerted from the intracellular surface of the cell membrane. The only known action of extracellular Mg²⁺ is the low-affinity inhibition of active transport of Ca²⁺ which is discussed in Chapter 6, Section I. Mg²⁺ cannot mimic the effects of Ca²⁺ on the Ca²⁺-ATPase and is not transported during the operation of the Ca²⁺ pump.[1]

A. Ca²⁺-ATPase Activities in the Absence of Added Mg²⁺

We have already mentioned in Chapter 8 that Ca²⁺-ATPase activity and Ca²⁺-dependent phosphorylation persist in media without added Mg²⁺. Under these conditions, ATPase activity is low and the low-affinity component of the substrate curve seems to be absent. The low-affinity component is probably completely dependent on added Mg²⁺.[2] Without added Mg²⁺, the rate of phosphorylation is low but the steady-state level of the phosphoenzyme is the same as in the presence of Mg²⁺. It has been mentioned in Chapter 8, Section I that the phosphoenzyme formed without Mg²⁺ dephosphorylates slowly and is not sensitive to acceleration by ATP.

It is not yet clear whether the enzymatic activities of the Ca²⁺ pump that are measurable in the absence of added Mg²⁺ imply that the Ca²⁺-ATPase can function in the total absence of Mg²⁺ or if they are the consequence of the stimulatory effect of endogenous Mg²⁺, either tightly bound to the enzyme preparation or a contaminant in the incubation media.

The participation of tightly bound Mg²⁺ is suggested by the observation of Muallem and Karlish[3] that preincubation and washing of red blood cell membranes in solutions containing 10 mM CDTA substantially increases the dependence of phosphorylation on added Mg²⁺. The possible role of contaminating Mg²⁺ has been analyzed in detail by Pershadsingh and McDonald.[4] These authors showed that in media with sufficiently high concentrations of CDTA, Ca²⁺-ATPase activity becomes completely dependent on exogenous Mg²⁺, an observation that has been confirmed by several authors.[5,6]

It must be pointed out, however, that the interpretation of the effects of CDTA as evidence in favor of the role of endogenous Mg²⁺ depends on the assumption that the only effect of the chelator is the removal of Mg²⁺, and this has not yet been proved.

B. The Kinetics of Activation by Mg²⁺

The effects of Mg²⁺ on Ca²⁺-ATPase activity may be exerted (1) through the formation of the MgATP complex, (2) by the direct binding of Mg²⁺ to a site in the enzyme, or (3) by a combination of both processes. It seems reasonable to think that if activation depended on the formation of MgATP, the kinetics of Mg²⁺ activation should be measured plotting activity vs. the concentration of MgATP, whereas if activation depended on the binding of Mg²⁺ to a site in the protein activities should be plotted against the concentration of Mg²⁺.

Although this reasoning is correct, it is easy to show that if certain conditions are fulfilled, activity will be a Michaelis-like function of the concentration of both MgATP or Mg²⁺, regardless of which of the ligands is the actual mediator of the activation by

Mg^{2+}. Since this is of practical interest as it allows one to some extent to study the kinetics of activation independently of its detailed mechanism, the conditions under which this property holds will be discussed.

1. Activity Vs. Mg²⁺ Concentration

If Mg^{2+} acted at a single class of noninteracting sites in the enzyme molecule, activation would be a hyperbolic function of Mg^{2+} concentration. To see to what extent this dependence is also manifested when the effects of the cation are mediated by MgATP, we can take the simple case of an enzyme that binds both free ATP and MgATP, but only catalyzes the hydrolysis of MgATP. Under rapid-equilibrium kinetics, the rate equation for such a system would be

$$v = \frac{Vm}{1 + \dfrac{Km}{[MgATP]}\left(1 + \dfrac{[ATPf]}{Ki}\right)} \tag{1}$$

where [ATPf] is the concentration of free ATP. Since

$$[ATPMg] = \frac{[ATPt]}{1 + \dfrac{Kd}{[Mg^{2+}]}} \tag{2}$$

where [ATPt] is the total ATP concentration and Kd the equilibrium constant for the dissociation of Mg^{2+} from MgATP, and

$$\frac{[ATPf]}{[MgATP]} = \frac{Kd}{[Mg^{2+}]} \tag{3}$$

Equation 1 can be rearranged to yield

$$v = \frac{Vm}{1 + \dfrac{Km}{[ATPt]} + \dfrac{KmKd}{[ATPt][Mg^{2+}]} + \dfrac{KmKd}{Ki[Mg^{2+}]}} \tag{4}$$

Equation 4 shows that when Mg^{2+} acts through the MgATP complex, if the total ATP concentration is kept constant, activity will increase with the concentration of Mg^{2+} along a Michaelis-like equation in which the maximum effect and the apparent affinity for Mg^{2+} will depend on the concentration of total ATP.

2. Activity Vs. MgATP Concentration

As is to be expected, hyperbolic responses to MgATP are predictable in the case that the complex mediates the effect of the cation. However, responses of this kind can also be observed when Mg^{2+} and ATP act at distinct and independent sites. This can be illustrated, taking the case of an enzyme with a site that does not discriminate between free ATP and MgATP and an independent site at which Mg^{2+} binds to act as an essential activator. For such a system, rapid-equilibrium considerations predict the following rate equation:

$$v = \frac{Vm}{\left(1 + \dfrac{Km}{[ATPt]}\right)\left(1 + \dfrac{Ka}{[Mg^{2+}]}\right)} \tag{5}$$

where Ka is the equilibrium constant for the dissocation of Mg^{2+} from the enzyme. Since according to Equation 2

$$[Mg^{2+}] = \frac{Kd}{\frac{[ATPt]}{[ATPMg]} + 1} \qquad (6)$$

Equation 5 can be rearranged to yield

$$v = \frac{Vm/(1 + Km/ATPt)}{1 + \frac{Ka[ATPt]}{Kd[MgATP]} - \frac{Ka}{Kd}} \qquad (7)$$

Equation 7 shows that if the total concentration of ATP is kept constant and the affinity of the enzyme for Mg^{2+} enzyme is greater than the affinity of ATP for Mg^{2+} (Kd > Ka), the activity will be a Michaelis-like function of the concentration of MgATP whose Km and Vmax will depend on the concentration of total ATP.

3. Activation by Mg²⁺ Under Steady-State Conditions

Let us consider the following steady-state reaction sequence that may be a part of a much more complicated cycle of reactions:

$$E + ATP \underset{k_{-1}}{\overset{k_1}{\rightleftharpoons}} EATP \overset{k_2}{\rightleftharpoons} etc. \qquad (8)$$

If neither the rate nor the equilibrium constant for the binding of ATP were modified by Mg^{2+}, but catalysis only proceeded in those molecules of EATP that were bound to Mg^{2+}, k_2 would be a function of Mg^{2+}. Assuming rapid-equilibrium binding of Mg^{2+} to a single class of noninteracting sites in EATP, the effect of Mg^{2+} would be formally expressible by making k_2 a hyperbolic function of Mg^{2+}, i.e.,

$$k'2 = \frac{k2}{1 + \frac{Ka}{[Mg^{2+}]}} \qquad (9)$$

Notice that since our reasoning is based on the assumption that Mg^{2+} does not affect ATP binding, thermodynamic considerations demand that Mg^{2+} should bind with identical effectiveness to the ATP in solution and to the ATP bound on the enzyme. Hence Equation 9 would be valid regardless of whether the site in EATP is provided by ATP or pertains to the enzyme. In this particular case, it may be impossible to decide on kinetic grounds alone if the effects of Mg^{2+} are mediated or not by the MgATP complex.

C. The Relation Between Ca²⁺-ATPase Activity and the Concentration of Mg²⁺

Before analyzing the mechanisms that have been proposed to account for the Mg^{2+}-requirement of the Ca^{2+}-ATPase, the experimentally observable effects of the cation on this enzymic activity will be described.

Figure 1 shows the result of an experiment in which the Ca^{2+}-ATPase activity of red blood cell membranes was measured in media containing different concentrations of

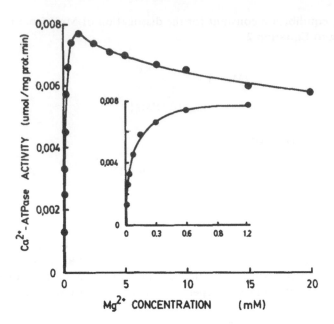

FIGURE 1.   Ca²⁺-ATPase activity from human red blood cell membranes as a function of Mg²⁺ in media with 20 μM ATP. The inset represents the activity up to 1.23 mM Mg²⁺. The continuous line was obtained by nonlinear regression and represents the solution of a kinetic equation that asumes that magnesium acts as activator with K_{Mg} = 0.088 mM, and as inhibitor with Ki = 9.64 mM and that Vm = 0.505 μmol/mg hr. Activity at 0 mM added MgCl₂ was 0.084 μmol/mg hr and the remaining activity when Mg²⁺ concentration tends to infinity 0.269 μmol/mg hr. The inset represents the initial part of the curve. (From Caride, A. J., unpublished.)

MgCl₂ and then plotted as a function of the concentration of Mg²⁺. It can be seen that the response of the ATPase to Mg²⁺ is biphasic: as the concentration of Mg²⁺ increases, activity first rises and after reaching a maximum at about 1.5 mM Mg²⁺, it progressively declines. The biphasic response to Mg²⁺ allows one to analyze the activity vs. Mg²⁺ concentration curve considering separately a high-affinity activating component and a low-affinity inhibitory component.

1. Activation by Mg²⁺

Activation of the ATPase by Mg²⁺ (inset in Figure 1) follows simple hyperbolic kinetics. The concentration of Mg²⁺ for half-maximal effect (K_{Mg}) is a linear function of positive intercept and slope of the concentration of Ca²⁺, indicating that Ca²⁺ is a competitive inhibitor of the activation by Mg²⁺.[7] Hence, to estimate genuine values of K_{Mg}, it is mandatory to extrapolate measured values to zero Ca²⁺ concentrations. When this is done, K_{Mg} becomes about 31 μM at low (20 μM) and about 53 μM at high (1 mM) ATP concentration.[7] These values are similar to those reported for the purified and solubilized enzyme.[8] Both at low and high concentrations of ATP, the Ki for Ca²⁺ is 30 μM, which is almost identical to K_{Mg}. This suggests that Ca²⁺ binds to the site at which Mg²⁺ binds and that this site is not selective for the latter cation. Therefore, specificity for activation by Mg²⁺ has to be attributed to the reactivity conferred by Mg²⁺ to the system rather than to the specificity of the binding of this cation. Neither K_{Mg} nor Ki for Ca²⁺ are modified by calmodulin, indicating that the site of the Ca²⁺-ATPase for Mg²⁺ activation is different than the site for Ca²⁺ activation.[7]

2. Inhibition by Mg²⁺

At Mg²⁺ concentrations above 2 mM, Ca²⁺-ATPase activity decreases. Analysis by nonlinear regression suggests that inhibition by Mg²⁺ takes place along a rectangular hyperbola which tends to a value of about 40% of the maximum activating effect of Mg²⁺ as Mg²⁺ concentration goes to infinity (Figure 1).[7] Inhibition is half-maximal at about 10 mM Mg²⁺. Neither changes in the concentration of Ca²⁺ nor calmodulin modify inhibition by Mg²⁺. The effect therefore can not be attributed to displacement of Ca²⁺.[7]

Since inhibition seems to be partial, it is probable that high concentrations of Mg²⁺, rather than blocking completely ATPase activity, drive the enzyme into a state of lower activity. This state may be that endowed with Ca²⁺ + ATP-dependent phosphatase activity, since, as has been mentioned in Chapter 8, Section IV, when ATPase activity is measured in the presence of phosphatase substrates, inhibition of the ATPase by Mg²⁺ seems to set in parallel with activation of the phosphatase by this cation.

D. The Mechanism of the Activation by Mg²⁺

1. MgATP as the Substrate

Since ATP is a good chelator of Mg²⁺ and the cytosolic concentration of Mg²⁺ is high, most of the intracellular ATP will be present as MgATP. This makes it reasonable to think that under physiological conditions, this complex is effective as substrate of the ATPase reaction. This raises the possibility that ATPase activity requires Mg²⁺, because the MgATP complex is the actual substrate of the system. In addition to what has been discussed before in connection with the possibility of MgATP as the substrate, it is important to take into account that experimental results from other cation-transport ATPases which also require Mg²⁺ for activity indicate that Mg²⁺ is not needed for the binding of ATP to the enzyme, since high-affinity binding of the nucleotide is detected even when all possible contaminating Mg²⁺ has been eliminated with chelators like EDTA.[9,10] Although no studies on equilibrium binding of ATP to the plasma membrane Ca²⁺-ATPase are available, it should not be surprising to learn that in this respect, the behavior of this enzyme is similar to that of other cation-transport ATPases. It would seem, therefore, that hypothetical mechanisms for the role of the MgATP complex have to be based on the reactivity of this complex for enzymatic catalysis and not on its affinity for the enzyme.

It has been mentioned in Chapter 6, Section III that the Ca²⁺-ATPase possesses two functionally distinct sites for ATP. Hence, detailed analysis of the role of the MgATP complex during ATPase activity are only possible if the interactions between Mg²⁺ and ATP are studied separately at each of these sites. The experimental results on the interactions between Mg²⁺ and ATP at the high-affinity site have been examined before. Taken together, these results seem to indicate that the high-affinity catalytic site of the Ca²⁺-ATPase does not discriminate between MgATP, CaATP, and free ATP and that any of the three species serves as substrate for phosphorylation.[3,11,12]

The interactions between ATP and Mg²⁺ at the low-affinity site for the nucleotide have been carefully studied by Muallem and Karlish[3] who measured the rate of the ATPase reaction in media containing high (100 to 2000 μM) ATP concentrations at different amounts of Mg²⁺ and Ca²⁺. Confrontation of the results with the estimated concentration of Mg²⁺, Ca²⁺, free-ATP, MgATP, and CaATP led these authors to postulate that MgATP is the effective ligand of the low-affinity site and that CaATP competitively inhibits the effect of MgATP (see also Sections II and III and Figure 10 of Chapter 6). As has been mentioned in Chapter 8, Section I, the requirement for MgATP at the low-affinity site of the ATPase is in apparent contradiction with the lack of requirement of MgATP during the acceleration of the hydrolysis of the phosphoenzyme by ATP, which is thought to be the cause of the low-affinity component of the substrate curve of the ATPase. No explanation for this puzzling contradiction is yet available.

2. Direct Binding of Mg²⁺ to the ATPase

Studies on the elementary steps of the ATPase reaction in Chapter 8, Section I show that Mg^{2+} increases the rate the phosphorylation and promotes the $E_1{\sim}P \rightarrow E_2{\sim}P$ transition. Any of the species of ATP seem to be effective substrate for phosphorylation, and the $E_1{\sim}P \rightarrow E_2{\sim}P$ transition is independent of ATP. This makes it likely that for the promotion of phosphorylation and of the $E_1{\sim}P \rightarrow E_2{\sim}P$ transition, rather than acting through the MgATP complex, Mg^{2+} binds directly to sites in the Ca^{2+} pump.

Since MgATP can be the substrate and may be essential for activation of the Ca^{2+} pump, it is obvious that the inhibition by Mg^{2+} shown in Figure 1 must be caused by direct interaction of Mg^{2+} with the enzyme, as interaction of Mg^{2+} with ATP would only lead to the formation of more of the MgATP complex.

Present knowledge on the activation of the Ca^{2+}-ATPase by Mg^{2+} can be summarized by saying that it requires binding of Mg^{2+} at high-affinity sites in the enzyme and to ATP. The relative importance of these two processes in the overall activation of the Ca^{2+} pump as well as the existence of a Mg^{2+}-independent activity of the Ca^{2+}-ATPase remain open questions that require further experimental analysis.

II. ALKALI METAL IONS

In 1970, we[14] showed that the (Ca^{2+} + ATP)-dependent phosphatase activity of red blood cell membranes was almost completely dependent on K^+. This finding was followed in 1971 by the report of Schatzmann and Rossi[15] that Na^+ or K^+ considerably increased the Ca^{2+}-ATPase activity of these membranes. These authors also studied the activation of the ATPase by Ca^{2+} in media with and without K^+ and concluded that the apparent affinity for Ca^{2+} of the monovalent cation-dependent ATPase was considerably lower (K_{Ca} 175 μM) than that of the monovalent cation-independent ATPase (K_{Ca} 4 μM). Activation of the red blood cell Ca^{2+}-ATPase by alkali metals was confirmed a few months later by Bond and Green[16] who found that Rb^+, but not Cs^+ or Li^+, could substitute for Na^+ or K^+. These authors, however, could not confirm Schatzmann and Rossi's finding that the monovalent cation-dependent ATPase had low apparent affinity for Ca^{2+}, but contrarily, reported that activation was not associated to any changes in K_{Ca}.

By the time of the above-mentioned findings, it was believed that alkali metals had no effect on the active fluxes of Ca^{2+}.[15] This is why initially the activation observed in the presence of alkali metal ions was considered to be the expression of an anomalous behavior of the Na^+-pump induced by Ca^{2+},[14,15] a damaged Ca^{2+} ATPase, or an ATPase unrelated to the Ca^{2+} pump.[16]

In 1973, we[17] provided experimental evidence that strongly suggested that the (Ca^{2+} + ATP)-dependent phosphatase activity of red blood cell membranes, which is highly dependent on K^+, was catalyzed by the Ca^{2+}-ATPase and showed, in agreement with the observation by Bond and Green, that there were no detectable differences between the apparent affinities for Ca^{2+} of the K^+-activated Ca^{2+}-ATPase and of the monovalent-cation-independent ATPase. In 1977, we[18] showed that the Ca^{2+}-ATPase possessed binding sites for Na^+ and K^+, since these cations protected the enzyme against inhibition by *N*-ethylmaleimide. The same year, Wolf et al.[19] attained the first successful solubilization and partial purification of the ATPase and demonstrated that the purified enzyme was still activated by alkali metal ions. Although these findings strongly suggested that stimulation by monovalent cations was a property of the Ca^{2+}-pumping ATPase, the physiological meaning of activation remained obscure, since it was believed that monovalent cations did not stimulate active Ca^{2+} transport.

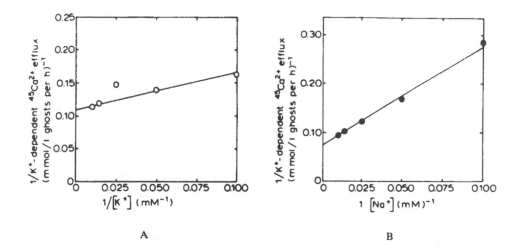

FIGURE 2.   Double reciprocal plots of K⁺-dependent ⁴⁵Ca efflux from human red blood cell ghosts as a function of K⁺ concentration (A), and Na⁺-dependent ⁴⁵Ca²⁺ efflux as a function of Na²⁺ concentration (B) on both sides of the membrane. K_K is 4.6 mM and K_{Na} is 24.1 mM. Maximum effect of K⁺ is 9.2 mmol/l ghosts hr and maximum effect of Na⁺ is 13.3 mmol/l ghosts hr. (From Kratje, R. B., Garrahan, P. J., and Rega, A. F., *Biochim. Biophys. Acta*, 731, 40, 1983. With permission.)

The discrepancy between the effects of alkali cations on ATPase and their lack of effect on active Ca²⁺ transport was removed in 1978 when Sarkadi et al.[20] demonstrated that the active uptake of Ca²⁺ into inside out vesicles of red blood cell membranes was activated by Na⁺ or K⁺ in the suspending media. The activation of Ca²⁺ transport by monovalent cations was confirmed by Wierichs and Bader[21] and by Romero[22] using resealed ghosts of red blood cells. These results definitively established that monovalent cations interacted directly with the Ca²⁺ pump of red blood cells to increase the rate of both Ca²⁺-ATPase and active Ca²⁺ transport.

No detailed studies of the effects of alkali cations on the Ca²⁺-ATPase activity of membranes from other cell types are yet available. In fact, it has been reported that 20 mM K⁺ has no effect on the Ca²⁺ ATPase of adipocyte membrane[23] or leads to slight inhibition on the ATPase of pancreatic islet membranes.[24] Therefore, it is not yet possible to determine whether activation by monovalent cations is a feature of all plasma membrane Ca²⁺-ATPases or a particular property of the red blood cell enzyme.

A. The Kinetics of Activation by Alkali Metal Ions

All studies in this respect have been performed measuring the Ca²⁺-ATPase activity and active Ca²⁺ transport in red blood cell membranes (Figure 2A,B).

Activity of this system is increased to the same extent by K⁺, Na⁺, Rb⁺, and NH₄⁺. Cs⁺ and Li⁺ have little or no effect.[19,25,26] Stimulation is not additive,[16] indicating that all cations act at the same class of site. Cardiac glycosides at concentrations that completely block (Na⁺ + K⁺) ATPase activity do not alter the activation.[15] Depending on the preparation and on the experimental conditions, at nonlimiting concentrations of monovalent cations, Ca²⁺-ATPase activity is 30 to 100% higher than in the absence of monovalent cations. Activation by alkali cations follows simple hyperbolic kinetics for both Ca²⁺-ATPase activity[15,16] and active Ca²⁺ transport.[25] At 37°C, $K_{0.5}$ for K⁺, NH₄⁺, or Rb⁺ is about 6 mM, and $K_{0.5}$ for Na⁺ is about 33 mM.[15,16,25] Hence the apparent affinity for Na activation is about 6 times lower than for the other monovalent cations. The apparent affinity is reduced considerably when ATPase is measured at 0°C ($K_{0.5}$ for K⁺ about 50 mM).[26] Calmodulin increases the maximum effect, the apparent affinity, and the selectivity for K⁺.[27] Activation by monovalent cations is associated with an increase in the maximum effect of Ca^{2+} and the Vmax for ATP without significant changes in the apparent affinity for Ca^{2+} or in the Km for ATP of the Ca^{2+}-ATPase;[26,27] hence there are no interactions among the apparent affinities of the site for monovalent cations and the sites for Ca^{2+} and ATP.

The phosphatase activity of the Ca^{2+} pump is much more dependent on alkali metal ions than the overall ATPase reaction. However, the apparent affinity for the stimulation of this activity is about five times less than that for stimulation of the ATPase although the K^+/Na^+ selectivity ratio is conserved.[17]

B. The Sideness of Activation

This problem has been studied measuring active Ca^{2+} efflux in resealed ghosts from human red blood cells. Romero[22,28] reported hyperbolic stimulation by external Na^+ or K^+ and suggested that the sites for activation faced the extracellular surface of the cell membrane. However, Romero's experiments also showed that replacement of intracellular Na^+ or K^+ by choline reduced the rate of Ca^{2+} transport, thus demonstrating an internal requirement for either cation.[28] The extracellular requirement for monovalent cations could not be confirmed in our laboratory.[25] In fact, using the preparation of resealed ghosts that is described in Chapter 6, Section I, we showed that the stimulation of Ca^{2+} efflux elicited by K^+ or by Na^+ at both surfaces of the cell membrane is preserved when choline replaces Na^+ or K^+ in the extracellular medium and is completely lost when choline replaces Na^+ or K^+ in the intracellular medium. It would seem, therefore, that the sites at which monovalent cations combine to stimulate the plasma membrane Ca^{2+} pump are accessible only from the inner surface of the cell membrane.

C. The Effects of Alkali Metal Ions on the Elementary Steps of the ATPase Reaction

This has been studied measuring the phosphorylation and dephosphorylation reactions of the Ca^{2+}-ATPase of red blood cell membranes[26] at 0°C.

Provided Ca^{2+} is present, K^+, Rb^+, NH_4^+, and Cs^+ increase the steady-state level of phosphorylation of the Ca^{2+} pump. This assertion is based on the finding that the extra phosphorylation requires Ca^{2+} and ATP with the same apparent affinity as the phosphorylation of the Ca^{2+} pump and the phosphoprotein formed coelectrophoreses with the phosphoenzyme of the Ca^{2+}-ATPase.[26]

As the concentration of K^+ tends to infinity the steady-state level of the phosphoenzyme tends to a value that is three to four times higher than the value observed in K^+-free media. This indicates that under conventional assay conditions, only a small fraction of the phosphate acceptor sites of the Ca^{2-} ATPase are phosphorylated. The meaning of this is discussed in length in Chapter 8, Section I.

When measured in media of identical composition and at the same temperature, steady-state phosphorylation and Ca^{2+}-ATPase activity increase with the concentration of K^+ along rectangular hyperbolae that are half-maximal at the same concentration of K^+ (Figure 3). The effect of K^+ on ATPase activity is, however, relatively higher than its effect on steady-state phosphorylation. As a consequence of this, the ratio Ca^{2+}-ATPase activity/phosphoenzyme level increases as K^+ concentration is raised. K^+ therefore stimulates the turnover of the phosphate in the phosphoenzyme. An effect of turnover is also indicated by the fact that K^+ decreases the half-time for phosphorylation and increases the first-order rate-constant for dephosphorylation.[26] The increase in turnover detected measuring the elementary steps of the ATPase reaction is also in agreement with the previously mentioned fact that K^+ stimulates overall ATPase activity, increasing its maximum rate without affecting its apparent affinity for Ca^{2+} or for ATP.

D. The Physiological Meaning of Activation

This remains obscure. In resealed ghosts of red blood cells containing high concen-

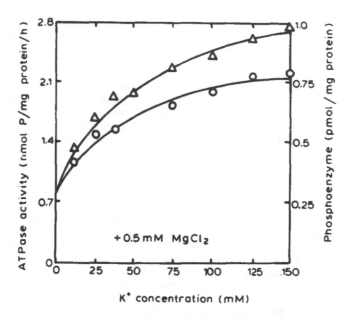

FIGURE 3.   The effect of K⁺ on Ca²⁺-dependent phosphoenzyme (O) and Ca²⁺-ATPase activity (Δ) of human red blood cell membranes measured at 0 to 3°C. (From Larocca, J. N., Rega, A. F., and Garrahan, P. J., *Biochim. Biophys. Acta*, 645, 10, 1981. With permission.)

trations of Na⁺, almost complete inhibition of Ca²⁺ efflux with vanadate is without effect on Na⁺ efflux.[25] This indicates that there is no correspondence between Ca²⁺ transport and Na⁺ efflux under conditions in which this cation activates the Ca²⁺ pump, suggesting rather strongly that during activation, Na⁺ is not transported. Since all cations act at the same class of sites, it seems reasonable to extend this finding to the rest of the activating cations and to conclude that activation is not the expression of the ability of the Ca²⁺ pump to transport monovalent cations.

Under physiological conditions, the cytosolic concentration of K⁺ is about 30 times higher than the $K_{0.5}$ for K⁺ activation of the Ca²⁺-ATPase. Therefore, changes in the concentration of K⁺ within the limits of cell viability will have little effect on Ca²⁺-ATPase. Hence, it is unlikely that K⁺ is a physiological modulator of this activity. However, since under physiological conditions the Ca²⁺-ATPase will be almost completely saturated by K⁺, the effect of K⁺ must be taken into consideration when performing experiments in vitro to analyze the physiological behavior of the Ca²⁺ pump.

REFERENCES

1. Schatzmann, H. J., Active calcium transport and Ca²⁺-activated ATPase in human red cells, *Curr. Topics Membr. Transp.*, 6, 125, 1975.
2. Richards, D. E., Rega, A. F., and Garrahan, P. J., Two classes of site for ATP in the Ca²⁺-ATPase from human red blood cell membranes, *Biochim. Biophys. Acta*, 511, 194, 1978.
3. Muallem, S. and Karlish, S. J. D., Studies on the mechanism of regulation of the red cell Ca²⁺ pump by calmodulin and ATP, *Biochim. Biophys. Acta*, 647, 73, 1981.
4. Pershadsingh, H. A. and McDonald, J. M., A high affinity calcium stimulated magnesium-dependent adenosine triphosphatase in rat adipocyte plasma membrane, *J. Biol. Chem.*, 255, 4087, 1980.
5. Verma, A. K. and Penniston, J. T.,. A high affinity Ca²⁺-stimulated and Mg²⁺-dependent ATPase in rat corpus luteum plasma membrane fractions, *J. Biol. Chem.*, 256, 1269, 1981.

6. Loterzstajn, S., Hanoune, J., and Pecker, F., A high affinity calcium-stimulated magnesium-dependent ATPase in rat liver plasma membranes: dependence on an endogenous protein activator distinct from calmodulin, *J. Biol. Chem.*, 256, 11209, 1981.

7. Caride, A. J., Garrahan, P. J., and Rega, A. F., unpublished observations.

8. Stieger, J. and Luterbacher, S., Some properties of the purified $(Ca^{2+} + Mg^{2+})$-ATPase from human red cell membranes, *Biochim. Biophys. Acta*, 641, 270, 1981.

9. Norby, J. G., Ligand interactions with the substrate site of Na, K-ATPase: nucleotides, vanadate and phosphorylation, *Curr. Topics Membr. Transp.*, 19, 281, 1983.

10. Dupont, Y. and Chapron, Y., Titration of the nucleotide binding sites of sarcoplasmic reticulum Ca^{2+} with 2',3'-0-(2,4,6-trinitrophenyl) adenosine 5'-triphosphate and 5'-diphosphate, *Biochim. Biophys. Res. Commun.*, 106, 1272, 1982.

11. Garrahan, P. J. and Rega, A. F., Activation of the partial reactions of the Ca^{2+}-ATPase from human red cells by Mg^{2+} and ATP, *Biochim. Biophys. Acta*, 513, 59, 1978.

12. Lichtner, R. and Wolf, H. U., Phosphorylation of the isolated high-affinity $(Ca^{2+} + Mg^{2+})$-ATPase of the human erythrocyte membrane, *Biochim. Biophys. Acta*, 598, 472, 1980.

13. Schatzmann, H. J.,. Role of magnesium in the $(Ca^{2+} Mg^{2+})$-stimulated membrane ATPase from human red cells, *J. Membrane Biol.*, 35, 113, 1977.

14. Garrahan, P. J., Pouchan, M. I., and Rega, A. F., Potassium-activated phosphatase from human red blood cells. The effects of adenosine triphosphate, *J. Membrane Biol.*, 3, 26, 1970.

15. Schatzmann, H. J. and Rossi, G. L., $Ca^{2+} + Mg^{2+}$-activated membrane ATPases in human red cells and their possible relations to cation transport, *Biochim. Biophys. Acta*, 241, 379, 1971.

16. Bond, G. H. and Green, J. W., Effects of monovalent cations on the $(Mg^{2+} + Ca^{2+})$-dependent ATPase of the red cell membrane, *Biochim. Biophys. Acta*, 241, 393, 1971.

17. Rega, A. F., Richards, D. E., and Garrahan, P. J.,. Calcium ion-dependent *p*-nitrophenyl phosphate phosphatase activity and calcium ion-dependent adenosine triphosphatase activity from human erythrocyte membranes, *Biochim. J.*, 136, 185, 1973.

18. Richards, D. E., Rega, A. F., and Garrahan, P. J., ATPase and phosphatase activities from human red cell membranes. I. The effects of N-ethylmaleimide, *J. Membrane Biol.*, 35, 113, 1977.

19. Wolf, H. U., Dieckvoss, G., and Lichtner, R., Purification and properties of high-affinity Ca^{2+}-ATPase of human erythrocyte membranes, *Acta Biol. Med. Ger.*, 36, 847, 1977.

20. Sarkadi, B., MacIntyre, J. D., and Gårdos, G., Kinetics of active calcium transport in inside-out red cell membrane vesicles, *FEBS Lett.*, 89, 78, 1978.

21. Wierichs, R. and Bader, H., Influence of monovalent ions on the activity of the $(Ca^{2+} + Mg^{2+})$-ATPase and Ca^{2+} transport of human red blood cells, *Biochim. Biophys. Acta*, 596, 325, 1980.

22. Romero, P. J., Active calcium transport in red cell ghosts resealed in dextran solutions, *Biochim. Biophys. Acta*, 649, 404, 1981.

23. Pershadsingh, H. A. and McDonald, J. M., A high-affinity calcium-stimulated magnesium-dependent adenosine triphosphatase in rat adipocyte plasma membrane, *J. Biol. Chem.*, 255, 4087, 1980.

24. Pershadsingh, H. A., McDaniel, M. L., Landt, M., Buy, C. G., Lacy, P. E., and McDonald, J., Ca^{2+}-activated ATPase and ATP-dependent calmodulin-stimulated Ca^{2+} transport in islet cell plasma membranes, *Nature*, 288, 492, 1980.

25. Kratje, R. B., Garrahan, P. J., and Rega, A. F., The effects of alkali metal ions on active Ca^{2+} transport in reconstituted ghosts from human red cells, *Biochim. Biophys. Acta*, 731, 40, 1983.

26. Larocca, J. M., Rega, A. F., and Garrahan, P. J., Phosphorylation and dephosphorylation of the Ca^{2+} pump of human red cells in the presence of monovalent cations, *Biochim. Biophys. Acta*, 645, 10, 1981.

27. Scharff, O., Stimulating effect of monovalent cations on activator-dissociated and activator-associated states of Ca^{2+}-ATPase in human erythrocytes, *Biochim. Biophys. Acta*, 512, 309, 1978.

28. Romero, P. J. and Romero, E., The affinity of the Ca^{2+} pump of human erythrocytes for external Na^+ or K^+, *Biochim. Biophys. Acta*, 691, 359, 1982.

Chapter 10

CALMODULIN AND OTHER PHYSIOLOGICAL REGULATORS OF THE Ca^{2+} PUMP

P. J. Garrahan

I. CALMODULIN

The discovery that the plasma membrane Ca^{2+} pump is regulated by calmodulin was the outcome of two observations that initially appeared to be unrelated. One was that the kinetic properties of the Ca^{2+}-ATPase depended on whether Ca^{2+} was present or not in the media used to isolate the membranes. The other was that the Ca^{2+}-ATPase of red blood cell membranes was activated by a soluble cytosolic protein.

In 1972, Scharff[1] reported that the specific activity of the Ca^{2+}-ATPase of red blood cell membranes isolated in media with Ca^{2+} is higher than that of membranes prepared in media containing EGTA. This was confirmed by Schatzmann,[2] who compared the kinetics of Ca^{2+} activation of the Ca^{2+}-ATPase of red blood cell membranes prepared in media with EDTA with that of membranes prepared in media with low concentrations of Ca^{2+}. In EDTA-treated membranes, activation by Ca^{2+} took place following a curve of lower apparent affinity than that of membranes prepared in Ca^{2+}-containing media. The low-affinity curve could not be fitted by a single hyperbola, indicating heterogeneity of the sites for the cation. Both Scharff and Schatzmann interpreted their results as indicating that either the presence of chelators or the removal of Ca^{2+} during hemolysis damaged the enzyme.

The idea of damage was soon disproved by further studies that showed that the effect of the chelators was reversible. This led Scharff and Foder[3] to propose that the Ca^{2+}-ATPase could be driven reversibly into two conformational states: state A is characteristic of membranes obtained in media with chelators. In this state, the enzyme has a low maximum velocity and a low apparent affinity for Ca^{2+}. State B is attained or induced if low concentrations (few $\mu mol/\ell$) of Ca^{2+} are present during the preparation of the membranes. In the B state, the ATPase has a high maximum velocity and a high apparent affinity for Ca^{2+}.

In 1972, Bond and Clough demonstrated that the red blood cell cytosol contained a soluble protein that activated the Ca^{2+}-ATPase[4] (Figure 1). This finding was confirmed and extended by several authors (see, for instance, References 5 and 8). In 1975, Quist and Roufogalis[7] produced experimental evidence showing that removal of a water soluble factor from red blood cell membranes decreased the apparent affinity to Ca^{2+} of the ATPase and that this effect was reversible.

Experiments by Farrance and Vincenzi,[9] Scharff and Foder,[3] and Hanahan et al.[10] established that chelators of Ca^{2+} drive the Ca^{2+}-ATPase into a low-affinity, low-activity state, because in the absence of Ca^{2+}, Bond and Clough's activator is removed from binding sites in the membrane. The deactivation process is prevented by Ca^{2+}, because in its presence, the activator remains attached; and it is reversed by Ca^{2+}, because when the cation is added, the activator binds again to the cell membrane.

In two communications that were published simultaneously in 1977, Gopinath and Vincenzi[11] and Jarret and Penniston[12] showed that purified calmodulin (at that time called "phosphodiesterase activator") had the same effect on the activity and on the apparent affinity for Ca^{2+} as the partially purified activator protein obtained from red blood cell cytosol. Furthermore, Jarret and Penniston[12] showed that calmodulin coelectrophoresed with the red blood cell activator and Gopinath and Vincenzi[11] demon-

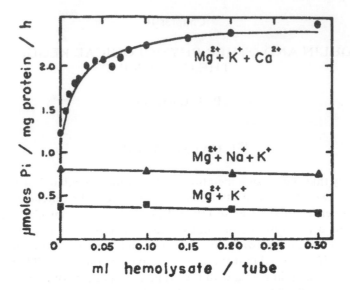

FIGURE 1.   The effect of a membrane-free hemolysate from human red cells on Mg²⁺, (Mg²⁺ + Na⁺ + K⁺)-, and (Mg²⁺ + Ca²⁺)-dependent ATPase activities of fragmented red cell membranes. (From Bond, G. H. and Clough, D. L., *Biochim. Biophys. Acta*, 323, 592, 1973. With permission.)

strated that other Ca²⁺-binding proteins (see Chapter 3, Section III) such as troponin C or parvalbumin were poor substitutes of calmodulin as activators of the ATPase.

In 1978, Jarret and Penniston[13] purified to homogeneity the red blood cell activator. This allowed detailed physical, chemical, and biological comparisons with calmodulin from other sources to establish definitely the identity of Bond and Clough's activator with calmodulin. It is now known that the effect of different preparation procedures on the kinetic properties of the Ca²⁺-ATPase activity results from the different amounts of endogenous calmodulin that these procedures leave attached to the isolated membranes.

A. Binding of Calmodulin to the Ca²⁺-ATPase

Calmodulin may affect an enzymatic activity either by binding to the enzyme or through the activation of a calmodulin-dependent protein kinase like those described in Chapter 3 which increase the rate of the Ca²⁺ pump of sarcoplasmic reticulum from heart muscle. The fact that activation by calmodulin persists after the Ca²⁺ pump has been separated from other membrane components by solubilization is a strong argument in favor of a direct interaction between calmodulin and the enzyme. However, the most compelling evidence for a direct interaction comes from quantitative studies of the binding of calmodulin to cell membranes and to the purified ATPase. Graf et al.[16] used calmodulin labeled with ¹²⁵I to show that in the presence, but not in the absence, of Ca²⁺, about 4000 binding sites with a dissociation constant of 1 to 5 nM could be detected in calmodulin-deficient red blood cell membranes. The number and affinity of these sites agreed with the estimations made in the same membranes[17] by studying the kinetics of calmodulin activation of the ATPase.

In a more recent study, Agre et al.[18] using functionally active ¹²⁵I-labeled calmodulin and red blood cell membranes detected high-affinity ($K_{0.5}$ = 0.3 to 0.5 nM) Ca²⁺-dependent binding sites which they identified as belonging to the Ca²⁺-ATPase (Figure 2). There were 1700 and 400 sites per cell measurable in isolated membranes and in inside out vesicles, respectively. These numbers are close to the number of copies of

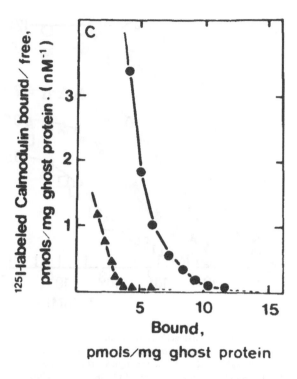

FIGURE 2.   Scratchard plot of ^{125}I-calmodulin binding to membranes (●) and IOVs (▲) from red cells. (From Agre, P., Gardner, K., and Bennet, V., *J. Biol. Chem.*, 258, 6258, 1983. With permission.)

the Ca^{2+}-ATPase per cell that can be estimated from the levels of Ca^{2+}-dependent phosphorylation. Agre et al.[18] also showed that the positive cooperativity for the binding to the calcium ATPase reported by several authors was an artifact caused by the incomplete equilibration of the binding reaction at very low calmodulin concentrations. Low-affinity, high-capacity calmodulin-binding sites were also detected and attributed, in agreement with previous findings by Sobue et al.,[19] to spectrin. Two membrane proteins that are extractable from red blood cell membranes in low-ionic strength solutions and have a M_r of 8000 and 40,000 also bind calmodulin with high affinity. The observations of Agre et al. strongly suggest that after treatment with chelators (which remove spectrin) and low-ionic strength solutions (which remove other loosely bound calmodulin-binding components), the Ca^{2+}-ATPase is the only calmodulin-binding protein that remains in red blood cell membranes. This is an important finding since, as mentioned in Chapter 5, Section II, the existence of a single calmodulin-binding species is a necessary condition for using calmodulin-affinity chromatography to purify this species.

All the studies of calmodulin binding we have mentioned insofar were performed on the assumption of a one to one stoichiometry for the binding of calmodulin to the Ca^{2+}-ATPase. A direct confirmation of this stoichiometry was provided by the experiments of Hinds and Andreasen.[20] These workers used azido-^{125}I-calmodulin which is a photoactivable radioactive derivative that retains the biological activity of calmodulin and is effective in specifically cross-linking with calmodulin-binding proteins in a Ca^{2+}-dependent fashion. Using both inside out vesicles and isolated membranes of red cells, Hinds and Andreasen showed that after treatment with the calmodulin derivative, the major product formed in both membrane types had an apparent M_r of 168,000 and that no other cross-linking product was common to both membrane types. The formation of the product of M_r 168,000 was Ca^{2+}-dependent and correlated well with the

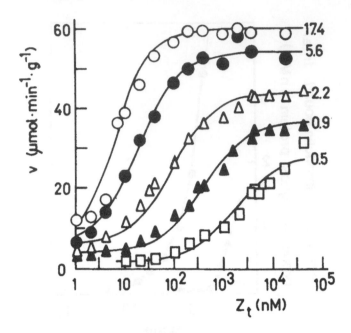

FIGURE 3.   Ca²⁺-ATPase activity from human red cell membranes
as a function of calmodulin concentration (Z_t) at the various Ca²⁺
concentrations (µM) indicated in the figure. (From Foder, B. and
Scharff, O., *Biochim. Biophys. Acta*, 649, 367, 1981. With permission.)

increase induced by azido calmodulin in active Ca²⁺-uptake into inside out vesicles and
in Ca²⁺-ATPase activity of isolated membranes. These results strongly suggest that the
product of M_r 168,000 dalton product represents a cross-link between calmodulin and
the Ca²⁺-ATPase. Hinds and Andreasen calculated a 1:1 M ratio for the cross-link
product, estimating that the M_r of the ATPase is 145,000 to 150,000 and that the
apparent M_r of calmodulin measured by SDS electrophoresis is 22,000.

Calmodulin binding to purified ATPase has been estimated from kinetic data[21] yielding results that can be quantitatively accounted for, assuming that calmodulin and the
ATPase bind in a 1:1 M ratio with a dissociation constant of 4.2 nM.

A final confirmation that calmodulin binds directly in a Ca²⁺-dependent fashion to
the Ca²⁺-ATPase is provided by the demonstration that if Ca²⁺ is present, the Ca²⁺-
ATPase of solubilized membranes is specifically retained in calmodulin-Sepharose columns, and is released from the columns when Ca²⁺ is removed (see Chapter 5, Section
II).

1. Role of Ca²⁺ in Calmodulin Binding

The binding of calmodulin to the Ca²⁺-ATPase is absolutely dependent on Ca²⁺. For
this reason, the apparent constant for the dissociation of calmodulin from the Ca²⁺-
ATPase is strongly affected by the concentration of Ca²⁺, binding decreasing about
1000 times as the concentration of Ca²⁺ lowers from 17.4 to 0.5 μM[22] (Figure 3). It is
likely that, as has been mentioned in Chapter 3, Section III for other calmodulin-
dependent enzymes, this dependence results from the fact that only the calmodulin-
Ca²⁺ complex has the adequate conformation to interact with the Ca²⁺-ATPase.

Several laboratories have addressed the question of which of the different complexes
of Ca²⁺ with calmodulin activate the ATPase. The experimental approach has been to
look for the best fits in plots of ATPase activity against the calculated concentration
of each of the Ca²⁺-calmodulin complexes. Although the agreement is not complete,

most of the experimental results seem to indicate that the complexes of calmodulin with three Ca^{2+} must be involved, whereas the complexes with two or with one Ca^{2+} are involved only slightly if at all.[23,24]

2. Extent of Calmodulin Dependence

In spite of the fact that the purified Ca^{2+}-ATPase is almost completely devoid of calmodulin, it catalyzes Ca^{2+}-dependent ATP hydrolysis. Furthermore, as will be discussed in this chapter, in the absence of calmodulin, either changes in the lipid environment or limited proteolysis can drive the Ca^{2+}-ATPase into a functional state that is almost identical to that attained when the enzyme is bound to calmodulin.

These results demonstrate that Ca^{2+} is a direct effector of the ATPase and that the sites at which Ca^{2+} binds to be transported pertain to the ATPase and are independent of calmodulin. In this respect, the Ca^{2+}-ATPase seems to differ from other calmodulin-dependent enzymes as those in Chapter 3 whose calcium-dependence can be fully accounted for by their dependence on the Ca^{2+}-calmodulin complex.

3. Binding of Calmodulin Under Physiological Conditions

The resting cytosolic Ca^{2+} concentration is less than that needed to saturate the Ca^{2+}-binding sites of calmodulin.[23] In spite of this, it has been argued that the large excess of calmodulin over Ca^{2+}-ATPase in red blood cells, and presumably in other cells, forces by mass action the binding of calmodulin to the ATPase and hence that: "for all practical purposes calmodulin is a subunit of the Ca^{2+} pump ATPase in vivo".[24] Present experimental evidence, however, suggests that this is not the case, since it has been shown by Foder and Scharff[22] that even when the concentration of calmodulin is very high, optimal calmodulin binding, as judged by optimal activation of the ATPase, requires concentrations of Ca^{2+} in the range of 1 to 10 μM, which is well above the physiological levels (Figure 3).

Additional information on the physiological regulation of the Ca^{2+} pump by calmodulin comes from studies of the rate constants for the binding of calmodulin to the ATPase. An estimate of these can be obtained from measurements of the rate of calmodulin-dependent activation and deactivation of the ATPase at different concentrations of Ca^{2+}. This procedure has been applied to red blood cell membranes by Scharff and Foder.[25] These workers adjusted their experimental data to a kinetic model that assumes that Ca^{2+} is in rapid equilibrium with Ca^{2+}-calmodulin and that the ATPase exists in two states, depending on whether it is combined or not with calmodulin. The estimated rate constant for calmodulin binding ranged from 2.5×10^5 ($M^{-1} \cdot min^{-1}$) at 0.7 μM Ca^{2+} to 1.73×10^7 ($M^{-1} \cdot min^{-1}$) at 10 μM Ca^{2+}. The estimated rate constant for the dissociation of calmodulin ranged from 6 (min^{-1}) at 0.1 μM Ca^{2+} to 0.044 (min^{-1}) at 2 to 20 μM Ca^{2+}. Using these values together with experimental values of equilibrium binding constants, Scharff and Foder calculated that at the resting cytosolic Ca^{2+} concentrations, no appreciable binding of calmodulin to the Ca^{2+}-ATPase took place. Their calculations also indicated that in stimulated cells, the large pool of cytosolic calmodulin would insure a rapid, but not instantaneous, response of the Ca^{2+}-ATPase to transient increases in the concentration of cytosolic Ca^{2+}.

The time-dependence of the activation of the Ca^{2+} pump by calmodulin measured by Scharff and Foder in isolated membranes predicts quantitatively the time course of the changes in red blood cell cytosolic Ca^{2+} induced by the Ca^{2+}-ionophore A23187,[26] and probably also explains the Ca^{2+}-induced oscillations in K^+ permeability (the "Gárdos effect" described in Chapter 3, Section VI) that is mediated by the ionophore in these cells.[27] A critical role for calmodulin binding and release in the in vivo regulation of the Ca^{2+} pump is also suggested by experiments on Ca^{2+}-dependent ATP hydrolysis in ATP-enriched intact red blood cells.[28]

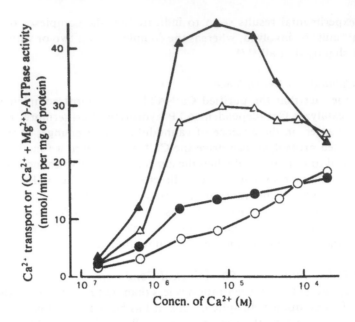

FIGURE 4.   Ca²⁺ uptake (solid symbols) and Ca²⁺-ATPase activity (open symbols) by IOVs of human red cells as a function of the concentration of Ca²⁺ in media with (triangles) and without (circles) calmodulin. (From Larsen, F. L., Katz, S., and Roufogalis, B. D., *Biochem. J.*, 200, 185, 1981. With permission.)

B. Effects of Calmodulin on the Steady-State Kinetics of the Ca²⁺-ATPase

The quantitative analysis of the effects of calmodulin in intact cell membranes is usually performed measuring the effects of added calmodulin on the activities of the Ca²⁺ pump. To do this, the following conditions must be fulfilled:

1. The membranes must be devoid of endogenous calmodulin. This is usually achieved by treating the membranes with Ca²⁺ chelators.
2. The Ca²⁺-ATPase must be protected against endogenous proteolysis, since partial proteolysis mimics the effects of calmodulin.
3. The membrane lipid fraction must be poor in acidic lipids, since these lipids mimic the effect of calmodulin.

It is likely that incomplete removal of endogenous calmodulin, partial proteolysis, and differences in the lipid environment explain the wide variation that exists in the literature concerning the magnitude of the effects of calmodulin on the Ca²⁺-ATPase of intact membranes.

An alternative way to quantify the effects of calmodulin is to use substances that selectively block the binding of calmodulin to the Ca²⁺ pump. The limitations and risks of this procedure are discussed in Chapter 11, Section III.

The most clear-cut effects of calmodulin on the overall ATPase reaction and on the active transport of Ca²⁺ are the increase in the apparent affinity for Ca²⁺ activation and the increase in the maximum rate of the reaction (Figure 8 of Chapter 6 and Figure 4).[23,29,41] In intact calmodulin-depleted membranes at physiological temperatures, calmodulin decreases K_{Ca} from more than 30 μM to less than 1 μM. Muallem and Karlish[32] have reported the interesting observation that at low temperatures (0°C), the apparent affinity for Ca²⁺ is very high, regardless of the presence or absence of calmodulin. Depending on the membrane preparation, calmodulin increases from two to nine times

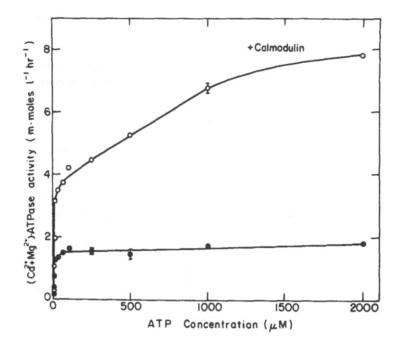

FIGURE 5.   Effect of calmodulin on Ca-ATPase activity from calmodulin-stripped membranes of human red cells at different ATP concentrations. (From Muallem, S. and Karlish, S. J. D., *Biochim. Biophys. Acta*, 597, 631, 1980. With permission.)

the maximum rate of ATPase activity and/or active Ca^{2+} transport. In purified ATPase, provided that the lipid environment does not mimic its effect, calmodulin induces an eight- to tenfold increase in activity and 20-fold increase in apparent affinity for Ca^{2+}.[33,35]

Although it is generally accepted that the effect of calmodulin on the rate of the ATPase is a direct effect on turnover, it has been argued by Schatzmann on the basis of experiments by Whutrich (Whutrich, unpublished, cited by Schatzmann[36]) that increases in the rate might be the consequence of the increased apparent affinity of the Ca^{2+} pump for Ca^{2+} which shifts the Ca^{2+}-activation curve away from the region of inhibitory Ca^{2+} concentration and thus allows a larger fraction of the activating potential of Ca^{2+} to be actually expressed. Schatzmann's argument requires calmodulin to increase the apparent affinity for the activation without change in the apparent affinity for inhibition by Ca^{2+}. This is difficult to reconcile with the findings by Muallem and Karlish[32] that at 0°C, calmodulin accelerates ATP hydrolysis without affecting the apparent affinity for Ca^{2+}.

Muallem and Karlish[31,32] showed that in calmodulin-stripped red blood cell membranes, the substrate curve of the Ca^{2+}-ATPase loses its low-affinity component and tends to a single Michaelis-Menten equation of high apparent affinity and low maximum velocity (Figure 5). The authors interpreted these results as showing that in the absence of calmodulin, the binding of ATP to the regulatory site in the ATPase does not occur, or if it does occur, it does not accelerate the turnover of the enzyme.

The interpretation of the effects of calmodulin on the substrate curve of the ATPase is complicated by the observation by Scharff[37] that the ATP-activation curve of calmodulin-deficient membranes is hyperbolic only at the low (30 μM) Ca^{2+} concentrations used by Muallem and Karlish. At higher (150 μM) Ca^{2+} concentrations, both calmodulin-deficient and calmodulin-saturated ATPases show biphasic substrate kinetics.

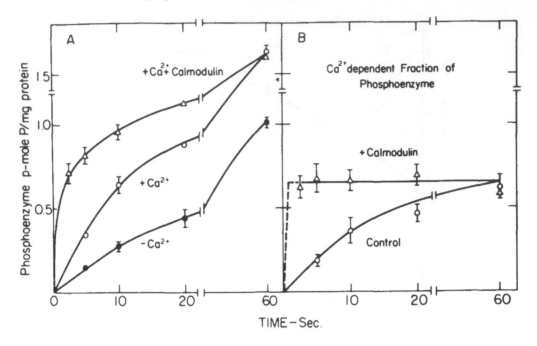

FIGURE 6.   Effect of calmodulin on the rate of phosphorylation of human red blood cell membranes by ³²P(ATP). (From Muallem, S. and Karlish, S. J. D., *Biochim. Biophys. Acta*, 597, 631, 1980.)

Calmodulin also modifies the activation of the ATPase by alkali metal ions. In calmodulin-saturated membranes, their activating effect is larger and is exerted with higher apparent affinity and higher specificity for K⁺ than in calmodulin-deficient membranes.[38]

C. Effects of Calmodulin on the Elementary Steps of the Ca²⁺-ATPase

All reported experiments on the effects of calmodulin on phosphorylation and dephosphorylation of the ATPase have been performed on red blood cell membranes and at 0°C. In media with Mg²⁺, calmodulin either has no effect or decreases the steady-state level of phosphorylation. Conversely, in the absence of added Mg²⁺, calmodulin increases the steady-state level of phosphorylation.[31,32,39-41] In both conditions, the ratio ATPase activity/phosphoenzyme level is increased,[39] which implies that calmodulin increases the turnover of the phosphoenzyme. In keeping with this effect, activation by calmodulin of the rate of phosphorylation (Figure 6) and of dephosphorylation has been reported.[31,32,40]

Although there seems to be consensus on these effects of calmodulin on the elementary steps of the ATPase reaction, our knowledge of these is still insufficient to postulate a molecular mechanism to account for the activation of the Ca²⁺-ATPase by calmodulin.

II. CONDITIONS AND TREATMENTS THAT MIMIC THE EFFECT OF CALMODULIN

In Chapter 7, Section I, we presented experimental evidence indicating that EGTA and some exogenous inorganic anions can substitute for calmodulin in its effects on the apparent affinity for activation by Ca²⁺. In this section, we will analyze how changes in the environment of the enzyme or modifications of its covalent structure give rise to responses that are similar to those elicited by calmodulin. Although it is unlikely that these phenomena play a role in the physiological regulation of the Ca^{2+} pump, they will be treated here, because they help in our understanding of the mechanisms of the effects of calmodulin.

A. The Lipid Environment

The enzymatic attack of the plasma membrane with phospholipases abolishes Ca^{2+}-ATPase activity.[42-44] The enzyme is not irreversibly damaged after treatment with phospholipases, since its activity can be restored by the addition of lipids.[43,44] In contrast with this, irreversible inactivation quickly sets in when the enzyme is solubilized with detergents. As has been mentioned in Chapter 5, inactivation is prevented if phospholipids are present in the solubilization media. These results suggest that the membrane lipids not only are needed for the functional competence of the ATPase, but also are essential for the structural integrity of the enzyme. The reason that inactivation is reversible after phospholipase treatment and irreversible after solubilization probably is that solubilization produces a more complete delipidation of the enzyme.

In 1980, Taverna and Hanahan[45] reported that when red blood cell membranes are treated with phospholipase A_2 under conditions in which there is little hydrolysis of membrane lipids, the Ca^{2+}-ATPase is activated much in the same way as when calmodulin is added. In 1982, Wetzker et al.[46] showed that oleic acid stimulated the red blood cell enzyme and inhibited competitively the binding of calmodulin. These results suggest that not only the functional and structural integrity of the Ca^{2+}-ATPase, but also its interaction with calmodulin depend on the lipid environment.

The availability of purified ATPase preparations, together with the development of procedures to reconstitute the enzyme into vesicles of controlled lipid composition, greatly facilitated the studies on the interaction of the Ca^{2+}-ATPase with its lipid environment. The main conclusion provided by these studies is that to preserve the structural and functional integrity of the enzyme, there is no specific requirement for a given class of membrane lipids, but that the effects of calmodulin are highly dependent on the physicochemical properties of the lipid environment. In fact, studies of the purified enzyme indicate that when reconstituted with acidic phospholipids such as phosphatidyl serine or polyunsaturated fatty acids, the ATPase acquires the high-affinity, high-activity state characteristic of the calmodulin-stimulated enzyme and loses its sensitivity to calmodulin. Full calmodulin sensitivity is observed when the enzyme is reconstituted with neutral phospholipids such as phosphatidylcholine[47,48] (Figure 7). It seems that the effects, but not the binding, of calmodulin are lost in the presence of acidic lipids, since soluble ATPase stabilized with phosphatidylserine is retained by calmodulin-Sepharose columns[49] with the same efficiency than when stabilized with phosphatidyl choline.[50]

Since acidic phospholipids are normal constituents of cell membranes, it is pertinent to ask to what extent their calmodulin-like effects are manifested under physiological conditions. This question was submitted to experimental test,[47,48] studying the calmodulin dependence of Ca^{2+}-ATPase purified from red blood cell membranes and reconstituted into liposomes containing different proportions of phosphatidylserine and phosphatidylcholine which are the main lipid components of red blood cell membranes. Half-maximal activation of the ATPase and, hence, half-maximal elimination of the calmodulin effect is reached when between 25 to 40% of the liposomal phospholipid is phosphatidylserine. Since only about 15% of the red blood cell membrane lipids is phosphatidylserine, these results indicate that in its native environment, the ATPase would be only partially activated by lipids. This is consistent with the significant calmodulin dependence experimentally observed in the Ca^{2+}-ATPase of red blood cell membranes.

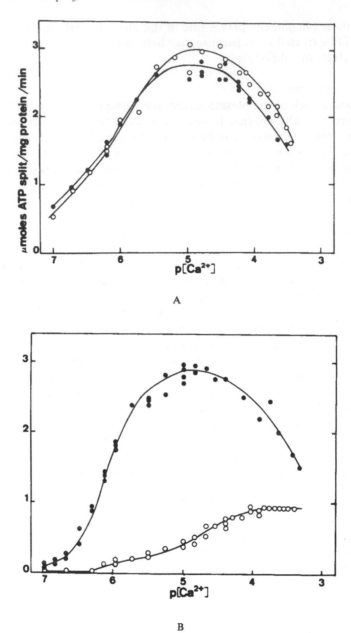

FIGURE 7.   (A) Ca²⁺ dependence of purified Ca²⁺ pump after reconstitution in phosphatidylserine liposomes in the presence (●) and in the absence (○) of calmodulin. (B) Ca²⁺ dependence of purified Ca²⁺ pump after reconstitution in phosphatidylcholine liposomes in the presence (●) and in the absence (○) of calmodulin. (From Niggli, V., Adunyah, E. S., Penniston, J. T., and Carafoli, E., *J. Biol. Chem.*, 256, 395, 1981.)

B. Proteolysis

Taverna and Hanahan[45] showed that mild treatment of isolated red blood cell membranes with trypsin or chymotrypsin increases the activity of the Ca²⁺-ATPase. This effect was studied in more detail by Sarkadi et al.[51] (Figure 8) and by Enyedi et al.[52] These authors, using IOVs of red blood cell membranes, demonstrated that mild trypsination reproduced the effects of calmodulin on the kinetic parameters of the Ca²⁺

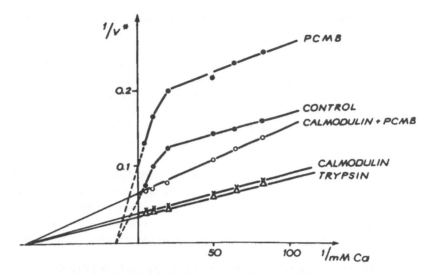

FIGURE 8.   Double reciprocal plot of the rate of Ca^{2+} uptake in mmol Ca^{2+}/mg protein min by IOVs from human red blood cells as a function of Ca^{2+} concentration. (From Sarkadi, B., Enyedi, A., and Gárdos, G., *Cell Calcium*, 1, 287, 1980.)

uptake and eliminated any further stimulation of Ca^{2+} uptake by calmodulin. Enyedi et al.[52] also showed that after trypsination, the M_r of the phosphoenzyme of the Ca^{2+}-ATPase is reduced by about 30,000. On the basis of these results, it was suggested that trypsin mimics the effects of calmodulin because it cleaves from the ATPase a regulatory domain of M_r near 30,000. The effects of proteolysis on the activity and on the calmodulin sensitivity of the Ca^{2+}-ATPase were confirmed by Stieger and Schatzmann[53] and by Carafoli et al.[48] using purified Ca^{2+}-ATPase. In addition, Rossi and Schatzmann[54] showed that when trypsin treatment was performed in the absence of Ca^{2+}, calmodulin sensitivity was lost without activation of the basal rate. Loss of calmodulin activation and calmodulin-like stimulation by trypsin require the presence of Ca^{2+} during treatment. The concentration of Ca^{2+} for half-maximal trypsin activation is equal to or less than the K_{Ca} of the Ca^{2+}-ATPase without calmodulin.

Enyedi et al.[52] and Agre et al.[18] have both shown that the high-affinity binding of calmodulin to red blood cell membranes is lost after treatment with trypsin, a result that supports the idea that the peptide that is released by proteolysis is the calmodulin-binding domain of the enzyme. However, Carafoli et al.[48] have shown that after proteolysis, azido-calmodulin labeled with ^{125}I remains covalently bound to the larger (M_r 90,000) tryptic fragment of the ATPase and that this fragment is retained by calmodulin-Sepharose columns indicating that the smaller peptide released by trypsination may not be the only part of the pump molecule which is involved in the binding of calmodulin.

C. The Mechanism of the Calmodulin-Like Effects of Acidic Lipids and Proteolysis

The effect of acidic lipids and proteolysis demonstrate that high activity and high apparent affinity for Ca^{2+} are not a unique property of the complex between calmodulin and the Ca^{2+}-ATPase. The similitude between the mechanism of action of calmodulin and of the procedures that mimic calmodulin is emphasized by the fact that some of the "anticalmodulin" drugs described in Chapter 11, Section III block the calmodulin-like effects of acidic lipids and proteolysis.[55] These effects are not unique for the Ca^{2+}-ATPase, since they have been observed in other calmodulin-dependent enzymes,[56,57] which suggests that calmodulin-like activation by acidic lipids and limited proteolysis may be a general property of these class of enzymes.

The fact that the effects of calmodulin can be reproduced by a series of very different maneuvers may be explained if we assume that at least two conformational states of different kinetic properties are accessible to the Ca²⁺ ATPase: a low-affinity, low-velocity and a high-affinity, high-velocity state. These states correspond closely to the A and B states proposed by Scharff.[3] Calmodulin, acidic lipids, and proteolysis would be equally effective in promoting the transition from the less to the more active state. Calmodulin and acidic lipids would do this by binding to a regulatory domain. Proteolysis would act, cleaving the regulatory domain, and thus freeing the rest of the molecule from the restraints that the regulatory domain imposes when it is not bound to calmodulin or to acidic amphiphiles.

Concerning the structure of the hypothetical regulatory domain, Gietzen et al.[58] have speculated that since the Ca²⁺-calmodulin complex is anionic and partially hydrophobic, for reasons of complementarity, the regulatory domain has to be hydrophobic and cationic. A domain with these properties would bind, in addition to Ca²⁺-calmodulin, anionic amphiphiles such as acidic phospholipids and unsaturated fatty acids.

III. OTHER PHYSIOLOGICAL REGULATORS

A. Protein Activators and Inhibitors

Maudlin and Roufogalis[59] reported the existence of a protein that activated the Ca²⁺-ATPase in a Ca²⁺-dependent fashion. The activator was extracted with EDTA from extensively washed erythrocyte membranes. Since it differed from calmodulin by a series of criteria, the authors concluded that it was a second modulator protein of the Ca²⁺-ATPase. This view has been contested by Au and Chan[60] who claim that the activator is in fact calmodulin and that the reported differences are a consequence of aggregation. In a recent study, Roufogalis et al.[61] reexamined the question and concluded that the membrane-bound activator was formed by calmodulin bound to a membrane protein that tightly binds calmodulin as a polymeric complex in a Ca²⁺-independent manner.

An endogenous protein activator distinct from calmodulin and a membrane-bound endogenous inhibitor of a supposedly plasma membrane Ca²⁺-ATPase of hepatocytes have been described.[62]

Au and Lee[64] and Whutrich[64] have reported the existence of inhibitor proteins in red blood cells extractable either from the plasma membrane[63] or from the cytosol.[64] The cytosolic inhibitor has been purified to apparent homogeneity, and it appears to be a single polypeptide of M_r 19,000 that inhibits specifically the Ca²⁺-ATPase by decreasing its apparent affinity for Ca²⁺ at all calmodulin concentrations.

B. Phosphoinositides

Phosphatidylinositol 4,5-biphosphate (P1-P2) is a powerful activator of red blood cell and synaptosomal Ca²⁺-ATPase.[48,66] The effect is exerted at relatively low membrane concentrations. The possible regulation of the Ca²⁺-ATPase by P1-P2 is interesting in view of the proposed involvement of phosphatidylinositol in Ca²⁺-mediated cell responses.

C. Regulation by Phosphorylation

There is some evidence that suggests that the Ca²⁺-ATPase of heart sarcolemma is regulated by cAMP-dependent phosphorylation.[67] Treatment of sarcolemmal vesicles with phosphorylase phosphatase, which dephosphorylates a number of sarcolemmal proteins, results in a reduction of both ATPase activity and active Ca²⁺ uptake. The inactivation is reversed by incubation in media with ATP, Ca²⁺, and Mg²⁺ and reactivation is blocked by an inhibitor of cAMP-dependent protein kinases. Moreover, the

ATPase reaction is activated by exogenously added phosphorylase B kinase. Since none of the above effects are observed in purified preparations of the sarcolemmal Ca^{2+}-ATPase, it has been proposed[68,69] that the heart sarcolemma contains a yet unidentified regulatory protein that is the target of a phosphorylation reaction that leads to activation of the ATPase. Regulation by phosphorylation seems to be unique to the heart sarcolemmal ATPase and has not been reported in Ca^{2+} pumps of other membranes. This may perhaps be the expression of the fact that heart activity is closely regulated by neurotransmitters and hormones which, in many cases, act through the modification of the Ca^{2+} fluxes. In this sense, it is suggestive that the heart muscle sarcoplasmic reticulum Ca^{2+}-ATPase is also regulated by phosphorylation.

REFERENCES

1. Scharff, O., The influence of calcium ions on the preparation of the (Ca^{2+}, Mg^{2+})-activated membrane ATPase in human red cells, *Scand. J. Clin. Lab. Invest.*, 39, 313, 1972.
2. Schatzmann, H. J., Dependence on calcium concentration and stoichiometry of the calcium pump in human red cells, *J. Physiol.*, 235, 551, 1973.
3. Scharff, O. and Foder, B., Low Ca^{2+} concentrations controlling two kinetic states of Ca^{2+}-ATPase from human erythrocytes, *Biochim. Biophys. Acta*, 483, 416, 1977.
4. Bond, G. H. and Clough, D. L., A soluble protein activator of $(Mg^{2+} + Ca^{2+})$-dependent ATPase in human red cell membranes, *Biochim. Biophys. Acta*, 323, 592, 1972.
5. Luthra, M. G., Hidenbrandt, G. R., and Hanahan, D. J., Studies on an activator of the $(Ca^{2+} + Mg^{2+})$-ATPase of human erythrocyte membranes, *Biochim. Biophys. Acta*, 419, 164, 1976.
6. Luthra, M. G., Hidenbrandt, G. R., Kim, H. D., and Hanahan, D. J., Observations on the $(Ca^{2+} + Mg^{2+})$-ATPase activator found in various mammalian erythrocytes, *Biochim. Biophys. Acta*, 419, 180, 1976.
7. Quist, E. E. and Roufogalis, B. D., Calcium transport in human erythrocytes: separation and reconstitution of high and low Ca affinity (Mg + Ca)-ATPase activities in membranes prepared at low ionic strength, *Arch. Biochem. Biophys.*, 168, 240, 1975.
8. Scharff, O., Ca^{2+} activation of membrane-bound $(Ca^{2+} + Mg^{2+})$ dependent ATPase from human erythrocytes prepared in the presence or absence of Ca^{2+}, *Biochim. Biophys. Acta*, 443, 206, 1976.
9. Farrance, M. L. and Vincenzi, F. F., Enhancement of $(Ca^{2+} + Mg^{2+})$-ATPase activity of human erythrocyte membranes by hemolysis in isoosmotic imidazole buffer. I. General properties of variously prepared membranes and the mechanism of the isoosmotic imidazole effect, *Biochim. Biophys. Acta*, 471, 49, 1977.
10. Hanahan, D., Taverna, R. D., Flynn, D. D., and Ekholm, J., The interaction of Ca^{2+}/Mg^{2+}-ATPase activator protein with human erythrocyte membranes, *Biochim. Biophys. Res. Commun.*, 84, 1009, 1978.
11. Gopinath, R. M. and Vincenzi, F. F., Phosphodiesterase protein activator mimics red blood cell cytoplasmatic activator of the $(Ca^{2+} + Mg^{2+})$-ATPase, *Biochim. Biophys. Res. Commun.*, 77, 1203, 1977.
12. Jarret, H. M. and Penniston, J. T., Partial purification of the $(Ca^{2+} + Mg^{2+})$-ATPase activator from human erythrocytes: its similarity to the activator of 3′:5′-cyclic nucleotide phosphodiesterase, *Biochim. Biophys. Res. Commun.*, 77, 1210, 1977.
13. Jarret, H. M. and Penniston, J. T., Purification of the Ca^{2+}-stimulated ATPase activator from human erythrocytes: its membership to the class of Ca^{2+}-binding modulator proteins, *J. Biol. Chem.*, 253, 4676, 1978.
14. Lynch, T. J. and Cheung, W. Y., Human erythrocyte $(Ca^{2+} - Mg^{2+})$-ATPase: mechanism of stimulation by Ca^{2+}, *Arch. Biochim. Biophys.*, 194, 165, 1979.
15. Niggli, V., Ronner, P., Carafoli, E., and Penniston, J. T., Effects of calmodulin on the $(Ca^{2+} - Mg^{2+})$-ATPase partially purified from erythrocyte membranes, *Arch. Biochim. Biophys.*, 198, 124, 1979.
16. Graf, E., Filoteo, A. G., and Penniston, J. T., Preparation of ^{125}I calmodulin with retention of full biological activity: its binding to human erythrocyte ghosts, *Arch. Biochim. Biophys.*, 203, 719, 1980.
17. Jarret, H. W. and Kyte, J., Human erythrocyte calmodulin. Further characterization and the site of its interaction with the membrane, *J. Biol. Chem.*, 254, 8237, 1979.
18. Agre, P., Gardner, K., and Bennett, V., Association between human erythrocyte calmodulin and the cytoplasmic surface of human erythrocyte membranes, *J. Biol. Chem.*, 258, 6258, 1983.

19. Sobue, K., Muramoto, Y., Fujita, M., and Kakiuchi, S., Calmodulin-binding protein of erythrocyte cytoskeleton, *Biochim. Biophys. Res. Commun.*, 100, 1063, 1981.

20. Hinds, T. R. and Andreasen, T. J., Photochemical cross-linking of axidocalmodulin to the (Ca^{2+} + Mg^{2+})-ATPase at low Ca^{2+} concentrations, *J. Biol. Chem.*, 256, 7877, 1981.

21. Graf, E. and Penniston, J. T., Equimolar interaction between calmodulin and the Ca^{2+}-ATPase from human erythrocyte membranes, *Arch. Biochim. Biophys.*, 210, 257, 1981.

22. Foder, B. and Scharff, O., Decrease of apparent calmodulin affinity of erythrocyte (Ca^{2+} + Mg^{2+})-ATPase at low Ca^{2+} concentrations, *Biochim. Biophys. Acta*, 649, 367, 1981.

23. Scharff, O., Calmodulin and its role in cellular activation, *Cell Calcium*, 2, 1, 1981.

24. Vincenzi, F. F., Hinds, T. R., and Raes, B. U., Calmodulin and the plasma membrane calcium pump, *Ann. N.Y. Acad. Sci.*, 356, 232, 1980.

25. Scharff, O. and Foder, B., Rate constants for calmodulin binding to Ca^{2+}-ATPase in erythrocyte membranes, *Biochim. Biophys. Acta*, 691, 133, 1982.

26. Scharff, O., Foder, B., and Skibstead, U., Hysteretic activation of the Ca^{2+} pump revealed by calcium transients in human red cells, *Biochim. Biophys. Acta*, 730, 295, 1983.

27. Vestergaard-Bogin, B. and Bennekou, P., Calcium-induced oscillations in K$^+$ conductance and membrane potential of human erythrocytes mediated by the ionophore A23187, *Biochim. Biophys. Acta*, 688, 37, 1982.

28. Muallem, S. and Karlish, S. J. D., Regulation of the Ca^{2+} pump by calmodulin in intact cells, *Biochim. Biophys. Acta*, 687, 329, 1982.

29. Larsen, F. L. and Vincenzi, F. F., Calcium transport across the plasma membrane: stimulation by calmodulin, *Science*, 204, 306, 1979.

30. Larsen, F. L., Katz, S., and Roufogalis, B. D., Calmodulin regulation of Ca^{2+} transport in human erythrocytes, *Biochim. J.*, 200, 185, 1981.

31. Muallem, S. and Karlish, S. J. D., Regulatory interaction between calmodulin and ATP in the red cell Ca^{2+} pump, *Biochim. Biophys. Acta*, 597, 631, 1980.

32. Muallem, S. and Karlish, S. J. D., Studies on the mechanism of regulation of the red cell Ca^{2+} pump by calmodulin and ATP, *Biochim. Biophys. Acta*, 647, 73, 1981.

33. Gietzen, K., Tejcka, M., and Wolf, H. U., Calmodulin affinity chromatography yields a functional purified erythrocyte (Ca^{2+} + Mg^{2+})-dependent adenosine triphosphatase, *Biochem. J.*, 189, 81, 1980.

34. Niggli, V., Adunyah, E. S., Penniston, J. T., and Carafoli, E., Purified (Ca^{2+} - Mg^{2+})-ATPase of the erythrocyte membrane. Reconstitution and effect of calmodulin and phospholipids, *J. Biol. Chem.*, 256, 395, 1981.

35. Stieger, J. and Luterbacher, S., Some properties of the purified (Ca^{2+} + Mg^{2+})-ATPase from human red cell membranes, *Biochim. Biophys. Acta*, 641, 270, 1981.

36. Schatzmann, H. J., The plasma membrane calcium pump of erythrocytes and other animal cells, in *Membrane Transport of Calcium*, Carafoli, E., Ed., Academic Press, London, 1982, 41.

37. Scharff, O., Kinetics of calmodulin-dependent (Ca^{2+} + Mg^{2+}) ATPase in plasma membrane and solubilized membranes from erythrocytes, *Arch. Biochim. Biophys.*, 209, 72, 1981.

38. Scharff, O., Stimulating effects of monovalent cations on activator dissociated and activator associated states of Ca^{2+}-ATPase in human erythrocytes, *Biochim. Biophys. Acta*, 512, 309, 1978.

39. Rega, A. F. and Garrahan, P. J., Effects of calmodulin on the phosphoenzyme of the Ca^{2+}-ATPase of human red cell membranes, *Biochim. Biophys. Acta*, 596, 487, 1980.

40. Luthra, M. G., Watts, R. P., Scherer, K. L., and Kim, H. D., Calmodulin. An activator of human erythrocyte (Ca^{2+} + Mg^{2+})-ATPase phosphorylation, *Biochim. Biophys. Acta*, 633, 299, 1980.

41. Jeffery, D., Roufogalis, B., and Katz, S., The effect of calmodulin on the phosphoprotein intermediate of Mg^{2+}-dependent Ca^{2+}-stimulated adenosine triphosphatase in human erythrocyte membranes, *Biochem. J.*, 194, 481, 1981.

42. Richards, D. E., Vidal, J. C., Garrahan, P. J., and Rega, A. F., ATPase and phosphatase activities from human red cell membranes. III. The effect of phospholipases on Ca-dependent enzymic activities, *J. Membr. Biol.*, 35, 137, 1977.

43. Roelofsen, B. and Schatzmann, H. J., The lipid requirement of the (Ca^{2+} + Mg^{2+})-ATPase in the human erythrocyte membranes, as studied by various highly purified phospholipases, *Biochim. Biophys. Acta*, 464, 17, 1977.

44. Ronner, P., Gazzotti, P., and Carafoli, E., A lipid requirement for the (Ca^{2+} + Mg^{2+})-activated ATPase of erythrocyte membranes, *Arch. Biochim. Biophys.*, 179, 578, 1977.

45. Taverna, R. D. and Hanahan, D. J., Modulation of human erythrocyte Ca^{2+}/Mg^{2+}-ATPase activity by phospholipase A2 and proteases. A comparison with calmodulin, *Biochim. Biophys. Res. Commun.*, 94, 652, 1980.

46. Wetzker, R., Klinger, R., and Frunder, H., Effects of fatty acids on activity and calmodulin binding of Ca^{2+}-ATPase of human erythrocyte membranes, *Biochim. Biophys. Acta*, 730, 196, 1983.

47. Niggli, V., Adunyah, E. S., Penniston, J. T., and Carafoli, E., Purified (Ca^{2+} + Mg^{2+})-ATPase of the erythrocyte membrane. Reconstitution and effect of calmodulin and phospholipids, *J. Biol. Chem.*, 256, 395, 1981.

48. Carafoli, E. and Zurini, M., The Ca²⁺-pumping ATPase of plasma membranes. Purification reconstitution and properties, *Biochim. Biophys. Acta,* 683, 279, 1982.

49. Niggli, V., Penniston, J. T., and Carafoli, E., Purification of (4Ca²⁺ + Mg²⁺)-ATPase from human erythrocyte membranes using a calmodulin affinity column, *J. Biol. Chem.,* 254, 9955, 1979.

50. Gietzen, K., Tejcka, M., and Wolf, H. U., Calmodulin affinity chromatography yields a functionally purified erythrocyte (Ca²⁺ + Mg²⁺)-dependent adenosine triphosphatase, *Biochem. J.,* 189, 81, 1980.

51. Sarkadi, B., Enyedi, A., and Gárdos, G., Molecular properties of the red cell calcium pump. I. Effects of calmodulin, proteolytic digestion and drugs on the kinetics of active calcium uptake in inside-out red cell membrane vesicles, *Cell Calcium,* 1, 287, 1980.

52. Enyedi, A., Sarkadi, B., Szász, I., Bot, B., and Gárdos, G., Molecular properties of the red cell calcium pump. II. Effects of calmodulin, proteolytic digestion and drugs on the calcium-induced membrane phosphorylation by ATP in inside-out red cell membrane vesicles, *Cell Calcium,* 1, 299, 1980.

53. Stieger, J. and Schatzmann, H. J., Metal requirements of the isolated red cell Ca²⁺ pump after elimination of calmodulin dependence by trypsin attack, *Cell Calcium,* 2, 601, 1981.

54. Rossi, J. P. F. C. and Schatzmann, H. J., Trypsin activation of the red cell Ca²⁺-pump ATPase is calcium-sensitive, *Cell Calcium,* 3, 583, 1982.

55. Adunyah, E. S., Niggli, V., and Carafoli, E., The anticalmodulin drugs trifluperazine and R24571 remove activation of the purified erythrocyte Ca²⁺-ATPase by acidic phospholipids and controlled proteolysis, *FEBS Lett.,* 143, 65, 1982.

56. Wolff, D. J. and Brostorm, C. O., Calcium-dependent cyclic nucleotide phosphodiesterase from brain: identification of phospholipids as calcium-independent activators, *Arch. Biochim. Biophys.,* 173, 720, 1976.

57. Depaoli-Roach, A. A., Gibbs, J. B., and Roach, P. J., Calcium and calmodulin activation of muscle phosphorylase kinase: effect of tryptic proteolysis, *FEBS Lett.,* 105, 321, 1979.

58. Gietzen, K., Sadorf, I., and Bader, H., A model for the regulation of the calmodulin-dependent enzymes erythrocyte Ca²⁺-transport ATPase and brain phosphodiesterase by activators and inhibitors, *Biochem. J.,* 207, 541, 1982.

59. Maudlin, D. and Roufogalis, B. D., A protein activator of Mg²⁺-dependent Ca²⁺-stimulated ATPase in human erythrocyte membranes distinct from calmodulin, *Biochem. J.,* 187, 507, 1980.

60. Au, K. S. and Chan, B. L., Nature of the (Ca²⁺ + Mg²⁺) ATPase activator protein which associates with the human erythrocyte membrane, *Biochim. Biophys. Acta,* 690, 261, 1982.

61. Roufogalis, B. D., Elliot, C. T., and Raiston, G. R., Characterization of a (Ca²⁺ + Mg²⁺)-ATPase activator bound to human erythrocyte membranes, *Cell Calcium,* 5, 77, 1984.

62. Lotersztajn, S., Hanoune, J., and Pecker, F., A high-affinity calcium-stimulated magnesium-dependent ATPase in rat liver plasma membranes. Dependence on an endogenous protein activator distinct from calmodulin, *J. Biol. Chem.,* 256, 11209, 1981.

63. Lotersztajn, S. and Pecker, F., A membrane-bound protein inhibitor of the high-affinity Ca-ATPase in rat liver plasma membranes, *J. Biol. Chem.,* 257, 6638, 1982.

64. Au, K. S. and Lee, K. S., An endogenous inhibitor of erythrocyte membrane (Ca²⁺ + Mg²⁺)-ATPase involved in calcium transport, *Int. J. Biochem.,* 11, 177, 1980.

65. Wuthrich, A., Isolation from hemolyzate of a proteinaceous inhibitor of the red cell Ca²⁺-pump ATPase. Its action on the kinetics of the enzyme, *Cell Calcium,* 3, 201, 1982.

66. Penniston, J. T., The plasma membrane Ca²⁺-pumping ATPases, *Ann. N.Y. Acad. Sci.,* 402, 296, 1982.

67. Caroni, P. and Carafoli, E., The Ca²⁺-pumping ATPase from heart sarcolemma. Characterization, calmodulin dependence and partial purification, *J. Biol. Chem.,* 256, 3263, 1981.

68. Caroni, P., Zurini, M., and Clark, A., The calcium-pumping ATPase of heart sarcolemma, *Ann. N.Y. Acad. Sci.,* 402, 402, 1982.

69. Caroni, P., Zurini, M., Clark, A., and Carafoli, E., Further characterization and reconstitution of the purified Ca²⁺-pumping ATPase of heart sarcolemma, *J. Biol. Chem.,* 258, 7305, 1983.

Chapter 11

INHIBITORS OF THE Ca²⁺ PUMP

P. J. Garrahan

I. INTRODUCTION

A number of very different kinds of substances are known to inhibit the plasma membrane Ca²⁺-ATPase. In spite of this, no selective and specific inhibitor of the Ca²⁺ pump, such as the cardiac glycosides in the case of the Na⁺ pump, has yet been found. A specific inhibitor would be useful to distinguish the Ca²⁺ pump activities from other unrelated Ca²⁺-dependent enzymic activities of the cell membrane.

Ca²⁺-ATPase activity is not affected by cardiac glycosides. This is a convenient property, since it provides an easy way to cancel any contribution of the (Na⁺, K⁺)-ATPase, which is always present in plasma membranes.

In this chapter, rather than undertaking a comprehensive survey of all the substances that inhibit the plasma membrane Ca²⁺ pump, those inhibitors that have yielded more information on the mechanism of the active transport reaction will be considered. They will be studied under four main headings namely: (1) inorganic ions; (2) calmodulin antagonists; (3) compounds that react with functional groups in proteins, and (4) other compounds which do not fit in the first three classes.

II. INORGANIC IONS

A. Lanthanides

Lanthanum (La³⁺) has been extensively used as an inhibitor of the Ca²⁺ pump. Most of our detailed knowledge of the effects of La³⁺ comes from studies in red blood cells.[1-7] Inhibitory effects of La³⁺ have also been reported in membrane Ca²⁺-ATPases from synaptic vesicles,[8,9] adipocytes[10] and renal cell plasma membranes.[11]

La³⁺ is inhibitory not only from the outside, but also from the inside of the red blood cell membrane. It has been reported[2] that extracellular La³⁺ reduces Ca²⁺-ATPase activity to a lesser extent than active Ca²⁺ transport (see Chapter 6, Section II), a finding that has been disputed.[6]

In isolated red blood cell membranes, inhibition by La³⁺ is exerted with a Ki of 2 to 3 μM and seems to be noncompetitive with respect to Ca²⁺ and ATP.[12] In red blood cells[4,13] and in basolateral membranes of kidney,[10] inhibition is accompanied by an increase in the steady-state level of the phosphoenzyme of the Ca²⁺-ATPase. Since in the presence of La²⁺ no net ATP hydrolysis takes place, it is reasonable to assume that the amount of phosphoenzyme formed in media with La²⁺ is close to the total capacity of Ca²⁺-dependent phosphorylation. Therefore, as shown in Chapter 8, the use of La³⁺ during phosphorylation may be a convenient procedure for estimating the total amount of Ca²⁺ pump units in a given preparation.

The mechanism of inhibition by La³⁺ at the level of the elementary steps of the ATPase reaction has been elucidated by Luterbacher and Schatzmann[13] and is described in Chapter 8, Section I. These authors demonstrated that La³⁺ increases the steady-state level of the phosphoenzyme, because it inhibits its hydrolysis. The rate of phosphorylation is not affected, and the phosphoenzyme formed in the presence of La³⁺ can be rapidly decomposed by reversing the phosphorylation reaction with ADP. When added after steady-state phosphorylation is reached in media containing Mg²⁺, La³⁺ is without effect on the stimulation of dephosphorylation by ATP + Mg²⁺. These effects are consistent with the idea that La^{3+} inhibits the Ca^{2+}-ATPase by preventing the Mg^{2+}-dependent $E_1P \rightarrow E_2P$ transition of the phosphoenzyme.

Little is known of the effect of lanthanides other than La^{3+}. Holmium (Ho^{3+}) and praseodymium (Pr^{3+}) inhibit both Ca^{2+}-ATPase and active Ca^{2+} transport in human red blood cells. Ki is about 10 to 20 μM. The effect is not specific for the Ca^{2+} pump, since (Na^+, K^+)-ATPase activity is also affected.[14] Ho^{2+}, Pr^{3+}, Gd^{3+}, (gadolinium) and Sm^{3+} (samarium) are as effective as La^{3+} in blocking active Ca^{2+} extrusion from Ca^{2+}-loaded intact red blood cells.[3]

B. Vanadate

The inhibitory effects of pentavalent vanadium (VO_3^-) were first observed in the (Na^+, K^+)-ATPase after the discovery that vanadate was present as an impurity in commercial ATP obtained from equine muscle.[15] Vanadium is present as an oligoelement in cells from most tissues. VO_3^- seems to inhibit most, if not all, cation-transport ATPases and is one of the most commonly used inhibitors of the Ca^{2+} pump of plasma membranes.

Under appropriate experimental conditions (see below), VO_3^- inhibits with high affinity (Ki = 5 μM) Ca^{2+}-ATPase activity and active Ca^{2+} transport in purified ATPase preparations,[16] in red blood cell membranes,[17,18] and in isolated plasma membranes from a wide variety of cells (see Chapter 4, Section II).

Detailed studies on the mechanism of VO_3^- inhibition of the Ca^{2+} pump in red blood cells have been performed by us[17,19] and by Bond and Hudgins.[18] Similar studies have been reported for the squid axon by DiPolo and Beaugé.[21,29] In red blood cells, VO_3^- decreases in parallel the overall Ca^{2+}-ATPase activity,[17,18] active Ca^{2+} transport,[19] and the steady-state level of phosphoenzyme.[17] The efficiency of VO_3^- as an inhibitor is strongly dependent on the ionic environment and on the concentration of ATP. This has to be taken into account when studying the sensitivity to vanadate of a given preparation. The anion is almost ineffective in the absence of added Mg^{2+} (Figure 1). K^+ increases inhibition in the presence of Mg^{2+}. Na^+, but not Li^+ replaces K^+ in this effect. The mechanism for the promotion of inhibition by Mg^{2+}, K^+, and Na^+ is the increase in the apparent affinity of the Ca^{2+}-ATPase for VO_3^-. In fact, addition of optimal concentrations of Mg^{2+} decreases the Ki for inhibition from more than 1 mM to about 25 μM. When K^+ is added in the presence of Mg^{2+} the Ki suffers a further drop to about 2 to 3 μM [17,18] (Figures 1 and 2).

The interactions in apparent affinity between VO_3^- and cations are linked so that the increase in apparent affinity for VO_3^- promoted by Mg^{2+}, K^+, or Na^+ is accompanied by a VO_3^--dependent increase in the apparent affinity for these cations as promoters of the inhibition. The apparent affinities for Mg^{2+}, K^+, and Na^+ for inhibition are lower than those for activation of Ca^{2+}-dependent ATP hydrolysis. As a consequence of this, the identification of the sites for inhibition with the physiological effector sites for the cations remains an open question.[17,18]

The effects of Mg^{2+} and K^+ on inhibition of the Ca^{2+}-ATPase are essentially similar to those reported for the (Na^+, K^+)-ATPase.[22,23] In this system also, the affinities of Mg^{2+} and K^+ for promotion of inhibition are less than those for stimulation of ATP hydrolysis. However, in the (Na^+, K^+)-ATPase, Na^+ antagonizes the effect of K^+.[22] It is likely that this is so, because in this system, Na^+ is not a congener of K^+ for activation, whereas in the Ca^{2+}-ATPase, Na^+ replaces K^+ in this respect.

Ca^{2+} at the low concentrations that are necessary for activation does not affect the apparent affinity for inhibition by VO_3^-, but at higher concentrations, Ca^{2+} progressively abolishes the inhibition by VO_3^-.[17]

Studies in reconstituted red blood cell ghosts[19] and in dialyzed squid axon[21] show that VO_3^- and the cations that promote inhibition are only effective at the inner surface

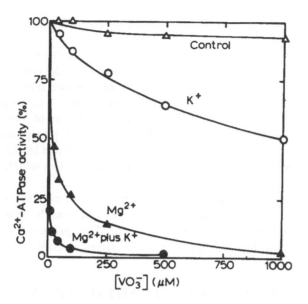

FIGURE 1. The effects of VO₃ on Ca²⁺-ATPase activity from human red cell membranes in media of different ionic composition. (From Barrabin, H., Garrahan, P. J., and Rega, A. F., *Biochim. Biophys. Acta,* 600, 796, 1980. With permission.)

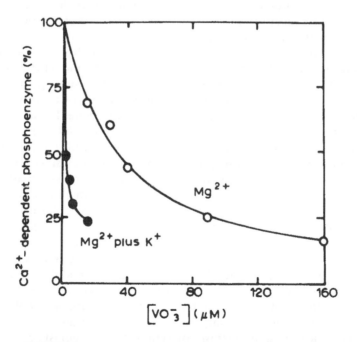

FIGURE 2. The effect of VO₃ on the steady-state level of the phosphoenzyme of the Ca²⁺-ATPase from red blood cell membranes in media with Mg²⁺ and with Mg²⁺ plus K⁺. (From Barrabin, H., Garrahan, P. J., and Rega, A. F., *Biochim. Biophys. Acta,* 600, 796, 1980. With permission.)

of the cell membranes, whereas release of inhibition by high Ca^{2+} concentration is only exerted by extracellular Ca^{2+}.

VO_3^- modifies in a complex way the substrate kinetics of the Ca^{2+}-ATPase.[17] On the high-affinity site, it acts as an noncompetitive inhibitor. At the low-affinity site, VO_3^- is a mixed-partially competitive inhibitor: the Km of this site increases hyperbolically with VO_3^- concentration, and even when ATP concentration is extrapolated to infinity, there persists an effect of VO_3^- (Figures 3 and 4). The effects of VO_3^- on the substrate kinetics of the Ca^{2+}-ATPase and the parallel decrease in phosphoenzyme level and ATPase activity can be predicted if we assume that VO_3^- binds to the E_2 conformer of the enzyme and prevents the $E_2 \rightarrow E_1$ transition, a mechanism which is similar to that proposed for the (Na^+, K^+)-ATPase.[24]

VO_3^- also inhibits the Ca^{2+}-ATPase from sarcoplasmic reticulum. The apparent affinity for inhibition in these systems seems to be significantly lower than that for inhibition of the plasma membrane Ca^{2+}-ATPase. This property has been used as a criterion for distinguishing plasma membrane from sarcoplasmic reticulum Ca^{2+}-ATPases.[25,26]

III. CALMODULIN ANTAGONISTS

Since the plasma membrane Ca^{2+} pump is not completely dependent on calmodulin, drugs that block the effect of calmodulin will be partial inhibitors of the pump activities. Inhibitors of calmodulin activation are potentially useful for studying regulation of the Ca^{2+} pump in complex systems in which procedures for stripping a membrane of endogenous calmodulin are not applicable. In many cases, however, this usefulness is hampered, because most of the calmodulin antagonists have additional effects apart from interfering with calmodulin activation.

The field of calmodulin antagonists was opened by the discovery of Weiss and co-workers[27] that the phenothiazine antipsychotic trifluoperazine (TFP) blocked Ca^{2+}-dependent phosphodiesterase activity. Levin and Weiss[28,29] subsequently showed that TFP inhibits such activity, because it binds to the calmodulin-Ca^{2+} complex (but not to free calmodulin) and impedes its interaction with the enzyme. It was soon found that other drugs acting on the central nervous system were calmodulin antagonists.[30,31] The attractive idea that calmodulin antagonism was the basis of the pharmacological effect of phenothiazines and other neuroleptics was soon disproved[32] since: (1) for a series of psychotropic agents, calmodulin antagonism correlates better with hydrophobicity (as judged by the water/octanol partition coefficient) than with clinical potency;[33] (2) antagonism with calmodulin shows no stereospecificity, whereas pharmacological effects of neuroleptics do;[32,33] (3) analogs of the phenothiazine chloropromazine differing in the position of the chlorine substitution in the aromatic ring are equally effective in antagonizing calmodulin activation, while only the 2-chloro analog has tranquilizer activity (Figure 5),[34] and (4) some of the most potent calmodulin antagonists known lack any pharmacological activity on the central nervous system (see below).

Most if not all the known calmodulin antagonists are cationic amphiphiles. On the basis of the structural complementary between calmodulin and its antagonists, it has been proposed as a general mechanism of action that antagonists bind to the hydrophobic region exposed by Ca^{2+} in calmodulin and competitively block its interaction with the target enzyme.[35]

A large and evergrowing number of substances have been identified as calmodulin antagonists, and a list of them has been compiled by Roufogalis,[32] including:

1.   Antipsychotics like the phenothiazines, the butophenones, the diphenylbutamines, and a miscellaneous group including clozapine and sulpiride

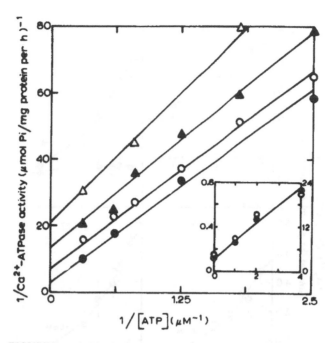

FIGURE 3. A Lineweaver-Burk plot of the Ca²⁺-ATPase activity of red cell membranes as a function of low (0.4 to 3.2 μM) ATP concentrations in media with 0 (●); 1 (○); 2 (△) and 4 (▲) μM VO₃⁻. The inset is the reciprocal of Km (○, μM,) left hand side ordinate) and Vm (●, μmol/mg protein/hr, right hand side ordinate) as a function of the concentration of VO₃⁻ (μM). (From Barrabin, H., Garrahan, P. J., and Rega, A. F., *Biochim. Biophys. Acta*, 600, 796, 1980. With permission.)

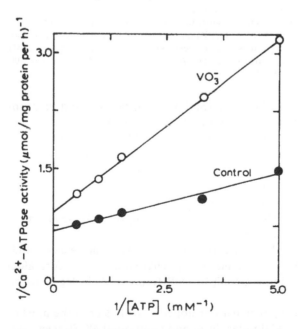

FIGURE 4. A Lineweaver-Burke plot of the low affinity component of the substrate curve of the Ca²⁺-ATPase vs. ATP concentration in the presence and absence of 1 μM VO₃⁻. (From Barrabin, H., Garrahan, P. J., and Rega, A. F., *Biochim. Biophys. Acta*, 600, 796, 1980. With permission.)

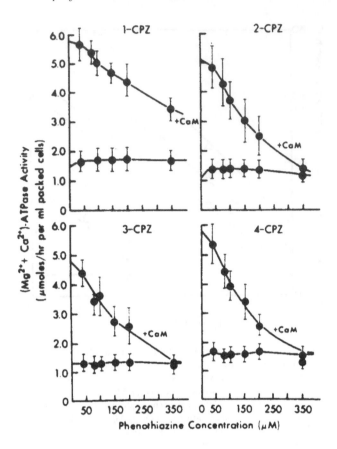

FIGURE 5.   Effect of chloropromazine (2-CPZ) and of chloropromazine analogs substituted with chlorine in position 1 (1-CPZ),
3 (3-CPZ) and 4 (4-CPZ) on the Ca²⁺-ATPase activity of red blood
cell membranes in the presence and absence of calmodulin (CaM).
(From Roufogalis, B. D., *Biochem. Biophys. Res. Commun.*, 98,
607, 1981. With permission.)

2.   Antidepressants like amitrylptyline, desipramine, and imipramine
3.   Muscle relaxants like compounds W7 to W10
4.   Minor tranquilizers like diazepam and chlordiazepoxide
5.   Local anesthetics like dibucaine, tetracaine, and lidocaine
6.   Antimitotic *Vinca* alkaloids like vinblastine and vincristine[40]
7.   *Rauwolfia* alkaloids
8.   Endogenous opiates such as endorphins
9.   Calmidazolium (formerly known as compound R24571), a derivative of the an-
      timycotic miconazole[41]
10.   Compound 48/80, which is used as a histamine releaser and is a condensation
      product of N-methyl-p-methoxyphenyltyamine with formaldehyde composed of
      a family of cationic amphiphiles differing in their degree of polymerization.[42]

Phenothiazines (in particular trifluoperazine) has been the most commonly used calmodulin antagonist. Calmidazolium and compound 48/80 seem to be the most potent,
since for both, Ki for inhibition is about 0.3 μM, a value at least one order of magnitude lower than the Ki value for the rest of the compounds mentioned in the preceding
paragraph.
Most of the calmodulin antagonists have other actions apart from blocking the effect

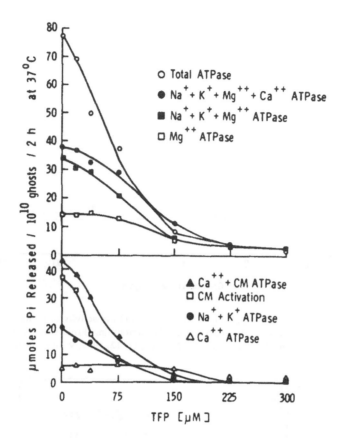

FIGURE 6. Effects of trifluoperazine (TFP) on the ATP-ase activities of rat red cell membranes (CM = calmodulin). (From Luttna, M. G., *Biochim. Biophys. Acta*, 692, 271, 1982. With permission.)

of calmodulin. In the case of the plasma membrane, Ca^{2+}-ATPase side effects include inhibition of the activity in the absence of calmodulin[43,44] and inhibition of the calmodulin-like effect of acidic lipids and of limited proteolysis.[45] Moreover, many calmodulin antagonists also inhibit calmodulin-independent membrane ATPases[43,46] (Figure 6) or other Ca^{2+} transport systems.[47] In general, side effects are exerted with lower apparent affinity than antagonism of calmodulin. The existence of side effects indicates that extreme caution has to be exerted, particularly in complex systems when interpreting the effects of calmodulin antagonists in terms of blockage of the effect of calmodulin on the plasma membrane Ca^{2+} pump.

Calmidazolium and compound 48/80 (Figure 7) seem to be the most selective[41,42,47] since their action on other ATPase activities or on the effect of acidic lipids or of proteolysis is either absent or exerted with considerably less apparent affinity than their effect on calmodulin activation. These compounds seem, therefore, to be the calmodulin antagonists of choice to use with the plasma membrane Ca^{2+}-ATPase.

IV. COMPOUNDS THAT REACT WITH PROTEINS

A. N-Ethylmaleimide (NEM)

In red blood cell membranes, NEM is an irreversible inhibitor of Ca^{2+}-dependent ATPase and phosphatase activities[48,49]: NEM also inhibits the (Na[+], K[+])-ATPase but, at least in red blood cells, the concentration of NEM required to reduce to one half

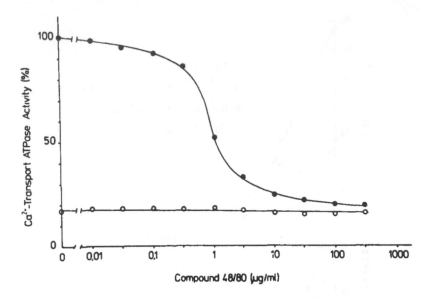

FIGURE 7. Effect of compound 48/80 on Ca²⁺-ATPase activity of human red cell membranes in media with (●) and without (○) calmodulin. (From Gietzen, K., Sánchez-Delgado, E., and Bader, H., *IRCS Med. Sci.*, 11, 12, 1983. With permission.)

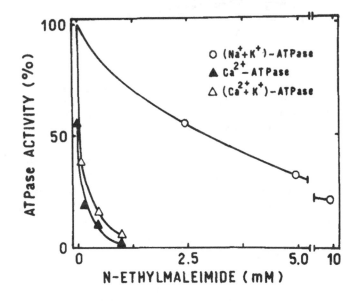

FIGURE 8. Effects of preincubation at 37°C during 30 min with different concentrations of *N*-ethylmaleimide on the ATPase activities of human red cell membranes. (From Richards, D. E., Rega, A. F., and Garrahan, P. J., *J. Membr. Biol.*, 35, 113, 1978. With permission.)

Ca²⁺ pump activities is about ten times less than necessary for the same effect on the Na⁺ pump[49] (Figure 8). The effect of NEM on the Ca²⁺ pump is modulated by physiological ligands. ATP, Na⁺, and K⁺ partially protect against inhibition, the effect of Na⁺ and K⁺ being less manifested than that of ATP. Conversely, Ca²⁺ increases from 5 to 10 times the sensitivity of the pump to NEM.[49] The effect of Ca²⁺ is exerted with high apparent affinity (K_{Ca} about 1 μM) and does not require ATP.[49] The lack of requirement of ATP for the interaction between NEM and Ca^{2+} is good proof that high-affinity binding of Ca^{2+} to the ATPase can take place in the absence of ATP.

B. Anion Channel (Band III) Inhibitors

The red blood cell membrane catalyzes a rapid passive exchange of intra- for extra-cellular anions. The system responsible for this has been identified with band III, the most abundant integral protein of red blood cell membranes.[50] Anion exchange is ir-reversibly inhibited by a series of reagents which include DIDS (4,4'-di-isothicyano-2,2'-stilbenes disulphonic acid) and the photolabeling reagent NAP-taurine (N-(4-azido-2-nitrophenyl)-2-aminoethyl sulfonate). In 1981, Waisman et al.[51] reported that active Ca^{2+} uptake into inside out red blood cell vesicles was significantly inhibited by NAP-taurine. As has been said in Chapter 6, Section I.E, the authors postulated that inhibition was caused by the blockage of the uptake of anions through band III which should accompany Ca^{2+} influx for reasons of electroneutrality. The inhibitory effect of NAP-taurine was confirmed by Minocherhomjee and Roufogalis.[52] These authors pre-sented evidence that inhibition has a direct effect on the Ca^{2+} pump since: (1) NAP-taurine inhibited Ca^{2+}-ATPase from the inner surface, whereas anion exchange is in-hibited from the outer surface of the cell membrane; (2) other anion channel blockers like probenecid are without effect on the Ca^{2+} pump; and (3) NAP-taurine inhibits purified Ca^{2+}-ATPase which lacks band III protein. In agreement with these findings, Niggli et al.[53] showed that NAP-taurine and DIDS inhibit purified ATPase in reconsti-tuted phospholipid vesicles in a fashion that is independent of charge-compensating mechanisms, since inhibition persists when Ca^{2+} accumulation into the vesicles is pre-vented by the ionophore A23187. It is likely that the discrepancy between the proposals of Waisman and co-workers and those of the other authors is only apparent and that depending on the experimental conditions, either blockage of anion channels or a di-rect effect on the Ca^{2+}-ATPase or both may cause inhibition of the Ca^{2+}-pump by DIDS or NAP-taurine. This is suggested by the observation that in intact red blood cells, band III inhibitors acting at the extracellular surface of the membrane markedly reduce Ca^{2+} transport without inhibiting Ca^{2+}-ATPase activity.[54]

C. Fluorescein Derivatives

These compounds inhibit cation-transport ATPases interacting with the sites for ATP.[55] This property was used by Karlish[56] and by Pick and Karlish[57] to develop fluo-rescein isothiocyanate (FITC) as an affinity ligand to covalently modify the sites for ATP of these ATPases. The effects of FITC on Ca^{2+}-ATPase activity and on Ca^{2+} uptake by inside out vesicles of red blood cell membranes have been studied.[58] The compound is an irreversible inhibitor of both phenomena, and inhibition is completely prevented by 3 mM ATP. The time course of the inhibition is biphasic, indicating the presence of both high- and low-affinity sites for ATP and FITC. Kinetic analysis of the inhibition indicates that the two classes of sites do not coexist, but pertain to alter-native conformational states of the pump protein. The studies with FITC have yielded information on the participation of different conformational states of the Ca^{2+}-ATP-ase in the elementary steps of the ATPase reaction and are further discussed in connec-tion with this subject in Chapter 8, Section I.

We[59] have analyzed the effect of the Erythrosin B (EB), a tetraiodinated derivative of fluorescein which binds in a noncovalent fashion to transport ATPases. EB com-pletely inhibits the Ca^{2+}-ATPase activity of red blood cell membranes. The effect is exerted reversibly at a single class of site(s) (Ki about 70 μM). EB acts as a noncompe-titive inhibitor at the catalytic site and as a competitive inhibitor at the regulatory site for ATP of the ATPase. These results are consistent with the idea that EB displaces ATP from the regulatory site without replacing it in its effects.

V. OTHER INHIBITORS

A. Quercetin

2 - [3,4-Dihydroxyphenyl](quercetin)-3,5,7-trihydroxy-4H-1-benzopyran-4-one is a flavonoid compound that inhibits with high affinity ($Ki \simeq$ 4 to 6 μM) both active Ca^{2+} uptake and Ca^{2+}-ATPase activity in inside out vesicles of red blood cells.[60] Inhibition in disrupted membranes takes place with an apparent affinity which is 5 times less than that of inside out vesicles. The reason for this difference is unknown and may be related to changes in the properties of the red blood cell membrane after vesiculization. The inhibitory effect is noncompetitive with respect to the activation of the Ca^{2+}-ATPase by Ca^{2+} or by ATP. The red blood cell membrane is highly permeable to quercetin, a fact which does not allow one to study the sideness of inhibition.[60] The action of quercetin is not specific for the Ca^{2+} pump, since the compound also inhibits the (Na^{+}, KI^{+})-ATPase[61] and the Ca^{2+}-ATPase of sarcoplasmic reticulum.[62]

B. Ruthenium Red

This is a cationic inorganic dye which has been useful in studying Ca^{2+} uptake by mitochondria. Ruthenium red inhibits Ca^{2+}-ATPase of red blood cell membranes.[63] This inhibitor has been superseded by newer and more effective compounds and practically is no longer used in the study of plasma membrane Ca^{2+}-ATPase.

REFERENCES

1. Weiner, M. L. and Lee, K. S., Active calcium ion uptake by inside-out and right-side out vesicles of red blood cell membranes, *J. Gen. Physiol.*, 59, 462, 1972.
2. Quist, E. E. and Roufogalis, B. D., Determination of the stoichiometry of the calcium pump in human erythrocytes using lanthanum as selective inhibitor, *FEBS Lett.*, 50, 135, 1975.
3. Sarkadi, B., Szász, I., Gerloczy, A., and Gárdos, G., Transport parameters and stoichiometry of active calcium ion extrusion in intact human red cells, *Biochim. Biophys. Acta*, 464, 93, 1977.
4. Schatzmann, H. J. and Burgin, H., Calcium in human red blood cells, *Ann. N.Y. Acad. Sci.*, 307, 125, 1978.
5. Szász, I., Sarkadi, B., Schubert, A., and Gárdos, G., Effects of lanthanum on calcium-dependent phenomena in human red cells, *Biochim. Biophys. Acta*, 512, 331, 1978.
6. Larsen, F. L., Hinds, T. R., and Vincenzi, F. F., On the red cell Ca^{2+}-pump: an estimate of stoichiometry, *J. Membr. Biol.*, 41, 361, 1978.
7. Szász, I., Hasitz, M., Sarkadi, B., Gárdos, G., Phosphorylation of the Ca^{2+} pump intermediate in intact red cells, isolated membranes and inside-out vesicles, *Molec. Cell Biochem.*, 22, 147, 1978.
8. Sorensen, R. G. and Mahler, H. R., Calcium-stimulated adenosine triphosphatases in synaptic membranes, *J. Neurochem.*, 37, 1407, 1981.
9. Michaelis, E. K., Michaelis, M. L., Chang, H. H., and Kitos, T., High-affinity Ca^{2+}-stimulated and microsomes, *J. Biol. Chem.*, 258, 6101, 1983.
10. de Smedt, H., Parys, J. B., Borghgraef, R., and Wuytack, F., Phosphorylated intermediates of (Ca^{2+} + Mg^{2+})-ATPase and alkaline phosphatase in renal plasma membranes, *Biochim. Biophys. Acta*, 728, 409, 1982.
11. Pershadsingh, H. J. and McDonald, J. M., A high-affinity calcium-stimulated magnesium-dependent adenosine triphosphatase in rat adipocyte membranes, *J. Biol. Chem.*, 255, 4087, 1980.
12. Rossi, J. P. F. C., Garrahan, P. J. and Rega, A. F., unpublished results.
13. Luterbacher, S. and Schatzmann, H. J., The site of action of La^{3+} in the reaction cycle of the human red cell membrane Ca^{2+}-pump ATPase, *Experientia*, 39, 311, 1983.
14. Schatzmann, H. J. and Tschabold, M., The lanthanides Ho^{2+} and Pr^{2+} as inhibitors of calcium transporting human red cells, *Experientia*, 27, 59, 1971.
15. Cantley, L. C., Josephson, L., Warner, R., Yanagisawa, M., Lechene, C., and Guidotti, G., Vanadate is potent (Na,K)-ATPase inhibitor found in ATP derived from muscle, *J. Biol. Chem.*, 252, 7421, 1977.

16. Niggli, V., Adunyah, E. S., Penniston, J. T., and Carafoli, E., Purified (Ca²⁺ − Ng²⁺)-ATPase of the erythrocyte membrane. Reconstitution and effect of calmodulin and phospholipids, *J. Biol. Chem.*, 256, 395, 1981.

17. Barrabin, H., Garrahan, P. J., and Rega, A. F., Vanadate inhibition of the Ca²⁺-ATPase from human red cell membranes, *Biochim. Biophys. Acta*, 600, 796, 1980.

18. Bond, G. H. and Hudgins, P., Inhibition of the red cell Ca²⁺-ATPase by vanadate, *Biochim. Biophys. Acta*, 600, 781, 1980.

19. Rossi, J. P. F. C., Garrahan, P. J., and Rega, A. F., Vanadate inhibition of active Ca²⁺ transport across human red cell membranes, *Biochim. Biophys. Acta*, 648, 145, 1981.

20. DiPolo, R., Rojas, H. R., and Beaugé, L., Vanadate inhibits uncoupled Ca efflux but not Na−Ca exchange in squid axons, *Nature*, 281, 228, 1979.

21. DiPolo, R. and Beaugé, L., The effects of vanadate on calcium transport in dialyzed squid axons. Sideness of vanadate-cation interaction, *Biochim. Biophys. Acta*, 645, 229, 1981.

22. Beaugé, L., Vanadate-potassium interaction on the inhibition of Na,K-ATPase, in *Na,K-ATPase: Structure and Function*, Skou, J. C. and Norby, J., Eds., Academic Press, New York, 1979, 373.

23. Bond, G. H. and Hudgins, P. M., Kinetics of inhibition of NaK-ATPase by Mg²⁺, K⁺, and vanadate, *Biochemistry*, 18, 325, 1979.

24. Karlish, S. J. D., Beaugé, L. A., and Glynn, I. M., Vanadate inhibits (Na⁺ + K⁺)-ATPase by blocking a conformational change of the unphosphorylated form, *Nature*, 282, 333, 1979.

25. Caroni, P. and Carafoli, E., The Ca²⁺-pumping ATPase of heart sarcolemma. Characterization, calmodulin dependence and partial purification, *J. Biol. Chem.*, 256, 3263, 1981.

26. Morcos, N. C., Localization of (Ca²⁺ + Mg²⁺)-ATPase, Ca²⁺ pump and other ATPase activities in cardiac sarcolemma, *Biochim. Biophys. Acta*, 688, 747, 1982.

27. Weiss, B., Fertel, R., Figlin, R., and Uzunov, P., Selective alteration of the activity of the multiple forms of adenosine 3′,5′-monophosphate phosphodiesterase of rat cerebrum, *Mol. Pharmaco.*, 10, 615, 1974.

28. Levin, R. M. and Weiss, B., Binding of trifluoperazine to the calcium-dependent activator of cyclic nucleotide phosphodiesterase, *Mol. Pharmacol.*, 13, 690, 1977.

29. Levin, R. M. and Weiss, B., Mechanism by which psychotropic drugs inhibit adenosine cyclic 3′,5′-monophosphate phosphodiesterase of brain, *Mol. Pharmacol.*, 12, 581, 1979.

30. Levin, R. M. and Weiss, B., Specificity of the binding of trifluoperazine to the calcium-dependent activator of phosphodiesterase and to a series of other calcium-binding proteins, *Biochim. Biophys. Acta*, 540, 197, 1978.

31. Levin, R. M. and Weiss, B., Selective binding of antipsychotics and other psychoactive agents to the calcium-dependent activator of cyclic nucleotide phosphodiesterase, *J. Pharmacol. Exp. Ther.*, 208, 454, 1979.

32. Roufogalis, B. D., Specificity of trifluoperazine and related phenotiazines for calcium-binding proteins, *Calcium and Cell Function*, 3, 129, 1982.

33. Norman, J. A., Drummond, A. H., and Moser, P., Inhibition of calcium-dependent regulator-stimulated phosphodiesterase by neuroleptic drugs is unrelated to their clinical potency, *Mol. Pharmacol.*, 16, 1089, 1979.

34. Roufogalis, B. D., Phenothiazine antagonism of calmodulin, a structurally-nonspecific interaction, *Biochim. Biophys. Res. Commun.*, 98, 607, 1981.

35. Gietzen, K., Sadorf, I., and Badr, H., A model for the regulation of the calmodulin-dependent enzymes erythrocyte Ca²⁺-transport ATPase and brain phosphodiesterase by activators and inhibitors, *Biochem. J.*, 207, 541, 1982.

36. Levin, R. M. and Weiss, B., Inhibition by trifluoperazine of calmodulin-induced activation of ATPase activity in rat erythrocytes, *Neuropharmacology*, 19, 169, 1981.

37. Raes, B. U. and Vincenzi, F. F., Calmodulin activation of red blood cell (Ca²⁺ + Mg²⁺)-ATPase and its antagonism by phenothiazines, *Mol. Pharmacol.*, 18, 253, 1980.

38. Hinds, T. R., Raess, B. U., and Vincenzi, F. F., Plasma membrane Ca²⁺ transport: antagonism by several potential inhibitors, *J. Membr. Biol.*, 58, 57, 1981.

39. Gietzen, K., Mansard, A., and Bader, H., Inhibition of human erythrocyte Ca⁺⁺-transport ATPase by phenothiazines and butyrophenones, *Biochem. Biophys. Res. Commun.*, 94, 674, 1980.

40. Gietzen, K., Wuthrich, A., and Bader, H., Effects of microtubular inhibitors on plasma membrane calmodulin-dependent Ca²⁺ transport ATPase, *Mol. Pharmacol.*, 22, 413, 1982.

41. Gietzen, K., Wuthrich, A., and Bader, H., R24571: a new powerful inhibitor of red blood cell Ca⁺⁺-transport ATPase and of calmodulin-regulated function, *Biochim. Biophys. Res. Commun.*, 101, 418, 1981.

42. Gietzen, K., Adamczyk-Engelmann, P., Wuthrich, A., Konstantinova, A., and Bader, H., Compound 49/80: selective and powerful inhibition of calmodulin-regulated function, *Mol. Pharmaco.*, in press.

43. Luthra, M., Trifluoperazine inhibition of calmodulin-sensitive Ca²⁺-ATPase and calmodulin insensitive (Na⁺ + K⁺) and Mg²⁺-ATPase activities of human rat red blood cells, *Biochim. Biophys. Acta*, 692, 271, 1982.

44. Vincenzi, F. F., Adunyah, E. S., Niggli, V., and Carafoli, E., Purified red blood cell Ca²⁺-pump ATPase: evidence for direct inhibition by presumed anti-calmodulin drugs in the absence of calmodulin, *Cell Calcium*, 3, 545, 1982.

45. Adunyah, E. S., Niggli, V., and Carafoli, E., The anticalmodulin drugs trifluoperazine and R24571 remove the activation of the purified erythrocyte Ca²⁺-ATPase by acidic phospholipids and by controlled proteolysis, *FEBS Lett.*, 143, 65, 1982.

46. Brody, T. M., Akera, T., Baskin, S. I., Gubitz, R., and Lee, C. Y., Interaction of Na,K-ATPase with chloropromazine free radical and related compounds, *Ann. N.Y. Acad. Sci.*, 242, 527, 1974.

47. Vale, M. G. P., Moreno, J. M., and Carvalho, A. P., Effects of calmodulin antagonists on the active Ca²⁺ uptake by rate liver mitochondria, *Biochem. J.*, 214, 929, 1984.

48. Bond, G. H., Ligand-induced conformational changes in the (Mg²⁺ − Ca²⁺)-dependent ATPase of red cell membranes, *Biochim. Biophys. Acta*, 288, 423, 1972.

49. Richards, D. E., Rega, A. F., and Garrahan, P. J., ATPase and phosphatase activities from human red cell membranes. I. The effects of N-ethylmaleimide, *J. Membr. Biol.*, 35, 113, 1977.

50. Cabantchik, Z. I., Knauf, P. A., Ostwald, T., Markus, R., Davidson, L., Breuer, W., and Rothstein, A., The interaction of an anionic photoreactive probe with the anion transport system of the human red blood cell, *Biochim. Biophys. Acta*, 455, 526, 1976.

51. Waisman, D. M., Gimble, J. M., Goodman, D. B. P., and Rasmussen, H., Studies of the Ca²⁺ transport mechanism of human erythrocyte inside-out vesicles. II. Stimulation of the Ca²⁺ pump by phosphate, *J. Biol. Chem.*, 256, 415, 1981.

52. Minocherhomjee, A. and Roufogalis, B. D., Selective antagonism of the Ca²⁺ transport ATPase of the red cell membrane by N-(4-azido-2-nitrophenyl)-2-aminoethylsulfonate (NAP-taurine), *J. Biol. Chem.*, 257, 5426, 1982.

53. Niggli, V., Sigel, E., and Carafoli, E., Inhibition of the purified and reconstituted calcium pump of erythrocyte by μM levels of DIDS and NAP-taurine, *FEBS Lett.*, 138, 164, 1982.

54. Waisman, D. M., Smallwood, J., Lafrenier, D., and Rasmussen, H., The role of band III in calcium transport across the human erythrocyte membrane, *FEBS Lett.*, 145, 337, 1982.

55. Skou, J. C. and Esmann, M., Eosin, a fluorescent probe of ATP binding to the (Na⁺ K⁺)-ATPase, *Biochim. Biophys. Acta*, 647, 232, 1981.

56. Karlish, S. J. D., Characterization of conformational changes in (Na,K)-ATPase labelled with fluorescein at the active site, *J. Bioenerg. Biomembr.*, 17, 111, 1980.

57. Pick, U. and Karlish, S. J. D., Regulation of conformational transition in the Ca-ATPase from sarcoplasmic reticulum by pH, temperature and calcium ions, *J. Biol. Chem.*, 257, 6120, 1982.

58. Muallem, S. and Karlish, S. J. D., Catalytic and regulatory ATP-binding sites of the red cell Ca²⁺ pump studied by irreversible modification with fluorescein isothiocyanate, *J. Biol. Chem.*, 258, 169, 1983.

59. Mugica, H., Rega, A. F., and Garrahan, P. J., The inhibition of the calcium dependent ATPase from human red cells by erythrosin B, *Acta Physiol. Pharmacol. Lat.*, 34, 163, 1984.

60. Wuthrich, A. and Schatzmann, H. J., Inhibition of the red cell calcium pump by quercetin, *Cell Calcium*, 1, 21, 1980.

61. Kuriki, Y. and Racker, E., Inhibition of (Na⁺, K⁺) adenosine triphosphatase and its partial reactions by quercetin, *Biochemistry*, 15, 4951, 1976.

62. Fewtrell, C. M. S. and Gomperts, B. D., Effect of flavone inhibitors of transport ATPase on histamine secretion from rat mast cells, *Nature*, 265, 635, 1977.

63. Watson, E. L., Vincenzi, F. F., and Davis, P. W., Ca²⁺-activated membrane ATPase: selective inhibition by ruthenium red, *Biochim. Biophys. Acta*, 249, 606, 1971.

INDEX

A

Access of calcium to binding site, 94
Acidic lipids, 93, 147—148
Acidic phospholipids, 77, 145
Actin-myosin association, 35
Activation
 of ATPase by magnesium, 130
 of calcium-ATPase by calcium, 77—78
 energy of for active transport of calcium, 97
 protein, 148
 substrate and, 84—85
Activation of calcium pump
 by alkali metal ions, 133—134
 by calcium, 71—72, 118
 by magnesium, 127, 129, 130
 physiological meaning of, 134—135
 sideness of, 134
Active sites
 in enzymes, 83—84
 for pNPP hydrolysis, 119
Acylphosphate, 108
Adenosine triphosphatase, see ATPase
Adenosine diphosphate, see ADP
Adenosine triphosphate, see ATP
Adenylate cyclase, 31
Adipocytes, 54, 56
ADP, 102
Aequorin, 3
Affinity for calcium, 74, 77, 91—95
 of calcium-binding proteins, 25
 modifiers of, 93
Affinity chromatography of calmodulin, 60—62
Alkali metal ions, see also specific metals, 132—135
Amino acids in calcium-ATPase, 62, 63
1,2-bis(o-Aminophenoxy)ethane-N,N,N′,N′-tetracetic acid (BAPTA), 3
Ammonium ion, 133
Amoeba, 10
Anions, see also specific types
 channel (band III) inhibitors of, 161
 organic, 93
 sites of, 6
Antagonists to calmodulin, 156—159
Antipyrylazo III, 2
Apparent affinity for calcium, 74, 77, 91—95
 modifiers of, 93
Arrhenius plot of calcium efflux, 97
Arsenazo III, 2
Artery muscle, 53
Artificial lipid membranes, 64
Aspartylphosphate, 108
ATP, 51, 114
 biphasic activation curves of, 82
 biphasic response to, 81, 82
 calcium-ATPase activity against concentration of, 81

cyclic, see Cyclic AMP
 dependence on, 80—87
 as energy-donating substrate, 81
 free, 87, 107, 113
 high-affinity catalytic site for, 114
 hydrolysis of, see ATP hydrolysis
 incorporation of phosphorus-32 into ADP to form, 102
 low-affinity noncatalytic site for, 114
 magnesium, see Magnesium-ATP
 phosphatase activity and, 119
 γ-phosphorus-32, 81
 reaction between pNPP sites and sites of, 119—122
 sites for, 113
 specific requirements for, 80
 as substrate, 81, 87
 synthesis of by calcium pump, 102
ATPase
 activation of by magnesium, 130
 amino acids in, 62
 binding of magnesium to, 132
 calcium-, see Calcium-ATPases
 calcium as direct effector of, 141
 dephosphorylation of, 144
 elementary steps of reaction of, 134
 magnesium-dependent activity of, 60
 maximum activity of, 77
 maximum velocity of, 56
 optimum pH for, 97
 phosphorylation of, 144
 sodium, potassium, 108, 120, 162
 sodium pump, 59
ATP hydrolysis, 77—80
 calcium dependence of, 106
 calcium transport and, 50
 coupling of calcium transport and, 50, 91—104
 dependence of on ligands, 114
 elementary steps of, 105—114
  dephosphorylation, 109—113
  phosphorylation, 105—110
 energy change during, 115—116
 during phosphatase activity, 119—122
 reaction scheme for, 114—115
Avena sativa (oats), 48
Axoplasm, 6
Azido-¹²⁵I-calmodulin, 139
N-(4-Azido-2-nitrophenyl)-2-aminoethyl sulfonate (NAP-taurine), 161

B

Band III (anion channel) inhibitors, 161
BAPTA, see 1,2-Bis(o-aminophenoxy)ethane-N,N,N′,N′-tetracetic acid
Barium, 91
Benzoic acid, 93

Binding
    calmodulin, see Calmodulin binding
    of magnesium to ATPase, 132
Binding of calcium, 27—29
    to calmodulin, 27—29
    by proteins, see Calcium-binding proteins
    site of, 94
Biphasic ATP activation curves, 82
Biphasic kinetics, 81
Biphasic response of calcium-ATPase
    to ATP, 81, 82
    to calcium, 78
Biphasic substrate curves, 83—86, 113
1,2-Bis(o-aminophenoxy)ethane-N,N,N′,N′-tetra-
    cetic acid (BAPTA), 3
2,7-Bisazo-1,8-dihydroxy-3,6-naphthalenedisul-
    fonic acid, 2
2,2-Bis(ethoxycarboxyl)methylamino-5-methyl-
    phenoxymethyl-6-methoxy-8-bis-(ethoxy-
    carbonyl)methylaminoquinoline (quin2), 3
Blockers of calcium channels, 23
Blood cells, red, see Red blood cells
Bone cells, 54, 56
Bound calcium, 6
Brain cells, 52, 56, 62

C

Calcineurin, 30, 32
Calcium
    cell function and, see also Cell function, 21—43
    cellular, 1—11
    homeostasis, 13—20
Calcium-activated photoproteins, 3
Calcium activation curves, 78
Calcium-ATP, 87, 107, 131
    as inhibitory ligand, 79
Calcium-ATPase, 56, 77
    in absence of added magnesium, 127
    activation by calcium, 77—78
    activity against ATP concentration, 81
    amino acids, 62, 63
    binding of calmodulin to, 138—141
    biphasic response to ATP, 81
    biphasic response to calcium, 78
    dependence on calcium, 78
    dephosphorylation, 109—113
    elementary steps of, 105—116, 144
    inhibition of, 154
    interaction with lipid environment, 145
    kinetics of, 142—144
    large-scale procedure for purification of, 61
    magnesium concentration and, 129—131
    partial reactions of, 105—125
    phosphatase activity of, 116—123
    phosphorylation, 105—110
    plasma membrane, 32—33
    purification of, 60—62
    purified, see Purified calcium-ATPase
    reaction mechanism of, 83

    substrate curve of, 143
    turnover of, 63
Calcium-binding dyes, 2—3
Calcium-binding proteins, 21, 25—29
    affinity of for calcium, 25
    vitamin D-dependent, 26
Calcium-dependence of ATP hydrolysis, 106
Calcium-dependent enzymes, 33—34
Calcium-dependent increase in potassium permea-
    bility, 37
Calcium-electron exchange, 13—15
Calcium-hydrogen exchange, 15, 76
Calcium-hydroxyl exchange, 76
Calcium-mediated cell functions, 21
Calcium-selective electrodes, 1, 3—5, 69
Calcium-sodium countertransport, 47
Calcium-sodium exchange, 13, 14
Caldesmon, 35, 36
Calmidazolium, 158
Calmodulin, 93, 94, 96, 97
    affinity chromatography of, 60—62
    alkali metal ions and, 133
    antagonists to, 156—159
    ATP hydrolysis and, 77, 78, 101
    binding of calcium to, 27—29
    binding to calcium-ATPase, 137—144
    cell types and, 52, 56
    chemical properties of, 27
    dependence of, 141
    dependence of enzymes on, 29—33
    dephosphorylation of ATPase, and 144
    distribution of, 27
    elementary steps of calcium-ATPase and, 144
    magnesium and, 130, 131
    phosphorylation of ATPase and, 144
    physical properties of, 27
    sensitivity to, 62
    steady-state kinetics of calcium-ATPase and,
        142—144
    transport of calcium and, 74, 77, 101, 142
    treatments that mimic effects of, 144—148
Calmodulin binding, 140—141, 147
    to calcium-ATPase, 138—141
    optimal, 141
    under physiological conditions, 141
    rate constants for, 141
    role of calcium in, 140—141
Calmodulin-binding domain, 147
Calmodulin kinase II, 32
Calmodulin-like effects, 145
Calmodulin-Sepharose conjugates, 60
Calpain, 26, 34
Calpastatin, 34
Carbohydrates, 62
Cardiac glycosides, 133, 153
Catalytic sites, 83
Cations, see also specific types
    divalent, 91
    monovalent, 118—119, 133, 134
Cell function, calcium and, 21—43
    binding proteins, 25—29

calmodulin and, 29—34
    messenger role, 21—24
    regulation, 34—39
Cell membrane phosphoinositides, 24
Cell membrane polyphosphoinositides, 24
Cells, see also specific types
    bone, 54, 56
    brain, 52, 56, 62
    calcium in, 1—10
    calcium-mediated functions of, 21
    channels of, 37—39
    circulating, 52
    communication between, 38—39
    distribution of calcium among components of,
        6—8
    division of, 36—37
    entrance of calcium into, see Influx of calcium
    epithelial, 9, 56
    Ehrlich ascites, 54—56
    excitable, 13—14, 22—23, 52—53
    exit of calcium from, see Efflux of calcium
    function of, see Cell function
    loading of with calcium, 67
    motility of, 35—36
    nonexcitable, 14—15
    optic nerve, 52
    pancreatic islet, 95
    potential of, 13
    protein kinases of, 32
    red blood, see Red blood cells
    regulation of functions of by calcium, 34—39
    resting, 17—18
    self-assembling of components of, 35—36
    stimulated, 21—24
    tissue, 53—54
Cellular calcium, 1—11
Cesium, 132
Channels
    blockers of, 23
    of calcium, 22—23
    cell-to-cell, 38
    membrane, 37—39
    potassium, 38
    surface density of, 23
    voltage-sensitive calcium, 13
Charge balance during calcium transport, 76
Chemical properties
    of calmodulin, 27
    of phosphoenzyme, 107—109
p-Chloromercurybenzene sulfonate (PCMBS), 67
Chloropromazine, 156
Cholate dialysis, 60, 70
Chromatography, 60—62
Chymotrypsin, 146
Circulating cells, 52
CNBr-Sepharose, 60
Communication between cells, 38—39
Composition of purified calcium-ATPase, 62—63
Compound 48/80, 158
Concentration of calcium
    in cytosol, 1—5, 13, 16—19

dependence on, 71—73, 77—80
    local, 5
    mean, 5
    ratio of, 13
    transients in, 5
Concentration of calmodulin in tissues, 27
Conductances, 23
Contractile proteins, 35
Contraction of muscle, 31
Corn (Zea mays), 48, 54, 56
Countertransport of calcium and sodium, 47
Coupling of calcium transport and ATP hydroly-
        sis, 50, 91—104
CTP, 80, 119
Cucurbita pepo (squash), 48
Currents of calcium, 23
Cyclic AMP, 21, 29, 30
    protein kinase dependent on, 32
    regulation of, 31
Cyclic GMP-dependent protein kinase, 32
Cyclic nucleotide metabolism enzymes, 31
Cyclic nucleotide phosphodiesterase, 31
Cytoplasmic calcium, 9
Cytoskeleton, 36
Cytosolic calcium, 7, 9—10
    concentration of, 1—5, 13, 16—19
    measurement of, 1—5
    regulation of, 16—19
    in stimulated cells, 21—24

D

Dependence
    on ATP, 80—87
    of calcium-ATPase on calcium, 78
    on calcium concentration, 71—73, 77—80
    of calmodulin, 141
    on ligands of elementary steps of ATP hydroly-
        sis, 114
    on magnesium, 117—118
    on pH, 96—97
    on temperature, 97—98
Dephosphorylation, 109—113, 116, 134, 144
    of ATPase, 144
    inhibition of, 79
Diacylglycerol, 24, 33, 34
Dialysis
    cholate, 60, 70
    internal, 70
DIDS, see 4,4-Diisothiocyano-2,2-stilbenedisul-
        fonic acid
Diffusion, 15
Digitonin, 5
Dihydropyridines, 23
4,4-Diisothiocyano-2,2-stilbenedisulfonic acid
    (DIDS), 75, 76, 161
Discovery of calcium pump, 45
Distribution
    of calcium, 1, 6—8
    of calmodulin, 27

Dithiothreitol, 74
Divalent-cation ionophore A23187, 67
Divalent cations, 91
Division of cells, 36—37
Dyes, see also specific dyes; specific types
    calcium-binding, 2—3
    tetracarboxylate, 2, 3

E

EDTA, 137
Efflux of calcium, 13—15
    Arrhenius plot of, 97
    curve for, 72
    from intact red blood cells, 67
    passive, 14
    by potassium, 134
    by sodium, 134
    uncoupled (residual), 14
Efflux of sodium, 135
EF-hand steric relationships, 25
EGTA, 3, 77, 94, 137
EGTA effect, 93—95
Ehrlich ascites cells, 54—56
Electrical balance during transport of calcium,
    75—79
Electrodes, calcium-selective, 1, 3—5, 69
Electrogenic transport of calcium, 75—76
Electroneutral transport of calcium, 76—77
Electron microprobe X-ray analysis, 6
Elementary steps
    of ATPase reaction, 134
    of ATP hydrolysis, 105—114
    of calcium-ATPase, 114, 144
    energy changes during, 115—116
Elevated intracellular calcium, 37
Endocytosis, 34—35
Endogenous magnesium, 127
Endoplasmic reticulum, 8, 15, 17, 24, 95
Energetics of calcium transport, 99
Energy changes during ATP hydrolysis, 115—116
Energy-donating substrate, 81
Enzymes, see also specific enzymes, 83
    active sites in, 83—84
    calcium-dependent, 33—34
    in cyclic nucleotide metabolism, 31
    dependence on calcium, 29—33
    dependence on calmodulin, 29—33
    in glycogen metabolism, 29—31
    in motile processes, 31
    purified, 61
    in regulation of muscle contraction, 31
    specific activity of purified, 61
Epithelial cells, 9, 56
Equilibrium constant, 116
N-Ethylmaleimide (NEM), 159—161
Exchange
    calcium-electron, 13—15
    calcium-hydrogen, 15, 76
    calcium-hydroxyl, 76

    calcium-sodium, 13, 14
Excitable cells, 13—14, 22—23, 52—53
Exocytosis, 34—35
External calcium, 79

F

Facilitated diffusion, 15
Fatty acids, 145
Fertilization, 35, 37
FITC, see Fluorescein isothiocyanate
Fluorescein derivatives, 161
Fluorescein isothiocyanate (FITC), 113, 161
Forty-eight/eighty compound, 158
Free ATP, 87, 107, 113
Free calcium, 1, 6
Freeze-thaw sonication procedure, 69

G

G actin, 26
Gadolinium, 154
Gap junctions, 38
Gardos effect, 37—38
Ghosts of red blood cells, 67—69, 71, 73
Giant axon, 9
Glutathione, 74
Glycogen, 29—31
Glycogen synthetase kinase, 30
Glycosides, 133, 153
GTP, see Guanosine triphosphate
Guanosine triphosphate (GTP), 80, 119

H

Heart muscle, 9
Heart sarcolemma, 53, 56, 62
Hemolysis, 68
Hepatocytes, see also specific types, 6, 8, 9, 19,
    148
High-affinity sites, 83, 87, 107, 114
Hill coefficient of calcium activation curves, 78
History, 45—51
Holmium, 154
Homeostasis of calcium, 13—20
Hydrogen-calcium exchange, 15, 76
Hydrogen translocation during calcium transport,
    76
Hydrolysis
    ATP, see ATP hydrolysis
    phosphoenzyme, 110, 112
    pNPP, 119
Hydroxylamine, 108
Hydroxyl-calcium exchange, 76
15-Hydroxyprostaglandin dehydrogenase, 33

I

Identification
  of calcium pump, 51
  of calcium transporting system, 49—51
Immunological cross reactivity, 64
Incorporation of phosphorus-32 into ADP to
    form ATP, 102
Indicators, metallochronic, 2
Influx of calcium, 13—15
Inhibition
  anion channel (band III), 161
  by calcium, 72—73, 78—79, 156
  of calcium-ATPase, 154
  of calcium pump, 153—162
  of calcium transport, 73, 74
  of dephosphorylation reaction, 79
  by high calcium, 156
  by intracellular calcium, 79
  by lanthanum, 112, 153
  by magnesium, 131
  of protein, 148
  release of, 156
Inhibitory ligand role of calcium-ATP, 79
Inorganic ions, see also specific types, 153—156
Inositol triphosphate, 19, 24
Inside-out vesicles (IOV), 67, 69, 75, 82, 87
  uptake of calcium by, 73, 94
Intact red blood cells, 67—68
Intercellular communication, 38—39
Internal dialysis, 70
  of squid axons, 47
Intestine, 53—54
Intracellular calcium, 1
  inhibition by, 79
  mitogenic response to elevated, 37
  release of calcium from, 24
Intracellular organelles, 6
  transport of calcium by, 15—16
Intravesicular calcium, 1
Ionized calcium, 1
Ionophore A23187, 67
Ionophores, see also specific types, 5
IOV, see Inside-out vesicles
Islet cells, 95
Isoelectric focusing, 60
Isolation of calcium pump, 59—65
ITP, 80, 119
  as substrate, 81

K

Kidney, 54, 56
Kinetics
  of activation by alkali metal ions, 133—134
  biphasic, 81
  of calcium-ATPase, 116—119, 142—144
  of calcium transport, 17—18
  equations for, 85—86

  of magnesium activation, 127—129
  of phosphorylation reaction, 106—107
  of purified calcium-ATPase, 63
  of substrate, 156
  of substrate curve, 83—86

L

Lanthanides, 153—154
Lanthanum, 52, 153
  inhibition by, 112, 153
Large-scale procedure for purification of calcium-
    ATPase, 61
Lectins, 37
Ligands, see also specific types
  dependence of elementary steps of ATP hydrol-
    ysis on, 114
  inhibitory, 79
Light scattering, 76
Lipid environment, 74, 145
Lipids, see also specific types
  acidic, 93, 147—148
  artificial membranes of, 64
Liposomes, 76, 101
  reconstituted, 67, 69—70
Lithium, 132
Liver microsomes, 14
Loading of cells with calcium, 67
Local calcium concentration, 5
Low-affinity sites, 83, 87, 113, 114, 119
  ATP at, 114, 131
  magnesium at, 131
Lymphocytes, 10, 52, 56

M

Magnesium, 7, 51, 114, 115
  activating effects of, 127
  activation by, 127—132
  ATPase activity dependent on, 60
  binding of to ATPase, 132
  calcium-ATPase and, 129—131, 154
  dependence on, 117—118
  endogenous, 127
  inhibition by, 131
  kinetics of activation by, 127—129
  at low-affinity site, 131
  phosphorylation, 106, 107
  potassium increase and, 93
Magnesium-ATP, 87, 107, 113, 127—129
  as substrate, 131—132
Manganese, 91
Mathematical equivalence of rate equations, 85—
    86
Maximum ATPase activity, 77
Maximum rate of calcium transport, 72, 74
Maximum velocity of calcium-ATPase, 56
Mean calcium concentration, 5
Measurement of calcium in cytosol, 1—5

Messenger role of calcium, 21—24
Metabolism
  cyclic nucleotide, 31
  of glycogen, 29—31
Metal ions, see also specific metals
  alkali, 132—135
Metallochronic indicators, 2
Microsomes, 14
Microtubules, 36
Mitochondria, 8, 15—19, 22, 24
  calcium storage and, 6
  uptake of calcium in, 15
Mitogenic response to elevated intracellular cal-
    cium, 37
Mitosis, 36
Mobilization of calcium, 24
Modifiers of apparent affinity for calcium, 93
Molecular weight of purified calcium-ATPase,
    62—63
Molybdate, 108
Monocytes, 52, 56
Monovalent cations, 118—119, 133, 134
Motile processes and enzymes, 31
Motility of cells, 35—36
Movement of calcium across plasma membrane,
    13
Muscle
  artery, 53
  cells of, 13
  heart, 9
  intestinal, 53
  regulation of contraction of, 31
  skeletal, see Skeletal muscle
  smooth, 6, 9, 56, 62
  stomach, 53, 56
  striated, 8
Mustard (*Sinapsis alba*), 48
Myosin-actin association, 35
Myosin light-chain kinase, 31, 35

N

NAD kinase, 33
NADP, 33
NAP-taurine, see *N*-(4-Azido-2-nitrophenyl)-2-
    aminoethyl sulfonate
NEM, see *N*-Ethylmaleimide
Nerve cells, 13, 17
Nerve terminals, 6, 8
Nervous tissue protein kinases, 32
Neurohypophysis, 52—53
Neurons, 9
Neurotransmitters, see also specific types, 32
  release of, 34
Neutral phospholipids, 145
Neutrophils, 52, 56
p-Nitrophenylphosphate, 116
Nonexcitable cells, 14—15
Null point titration, 5
Number of calcium sites, 95—96

O

Oats (*Avena sativa*), 48
Oleic acid, 145
Oocytes, 10
Optic nerve cells, 52
Optimal calmodulin binding, 141
Optimum pH, 97
Organelles
  calcium in, 6
  intracellular, see Intracellular organelles
Organic anions, see also specific types, 93
Oxalate, 69

P

Pancreas, 54, 56
Pancreatic islet cells, 95
Partial proteolysis, 77
Partial reactions of calcium-ATPase, 105—125
  elementary steps of ATP hydrolysis, 105—116
  phosphatase activity, 116—123
Parvalbumin, 25, 26, 138
Passive efflux, 14
Passive permeability, 14
PCMBS, see *p*-Chloromercurybenzene sulfonate
Permeability
  passive, 14
  potassium, 37
pH, 96—97
Phenothiazine, 28, 33, 38, 52, 156
Phenothiazine-Sepharose, 29
Phosphatase, 131
  active calcium transport and, 122—123
  ATP and, 119
  ATP hydrolysis during activity of, 119—122
  calcium-ATPase and, 116—123
Phosphate, 6, 62
Phosphatidylcholine, 61, 145
Phosphatidylinositol, 33, 148
Phosphatidylinositol, 4,5-bis-phosphate (PIP2),
    24, 48
Phosphatidylinositol 4-phosphate (PIP), 24
Phosphatidylserine, 60, 145
Phospodiesterase, 31
Phosphoenzyme, 105, 106, 134, 153
  chemical properties of, 107—109
  hydrolysis of, 110, 112
  turnover of, 144
Phosphoinositides (PI), 24, 148
Phosphokinase C, 24, 26
Phosphalamban kinase, 31
Phospholipases, 120, 145
  platelet, 33
Phospholipids, 6
  acidic, 77, 145
  neutral, 145
Phosphorus-32-ATP, 81

Phosphorus-32 incorporation into ADP to form
    ATP, 102
Phosphorylase kinase, 26
    skeletal muscle, 29—30
Phosphorylase phosphatase, 148
Phosphorylation, 105—109, 116, 134
    of ATPase, 144
    kinetics of, 106—107
    regulation by, 148—149
    reversal of, 107
    steady-state level of, 107
Photoproteins, 3
Physical properties of calmodulin, 27
Physiological conditions and binding of calmodu-
    lin, 141
Physiological meaning of activation, 134—135
Physiological regulators of calcium pump, 137—
    149
PI, see Phosphoinositides
PIP, see Phosphatidylinositol 4-phosphate
PIP2, see Phosphatidylinositol 4,5-bis-phosphate
Platelets, 10, 33
pNPP, 119—122
Poly(L-aspartic acid), 93
Poly(L-glutamic acid), 93
Polyphosphoinositides, 24
Polyunsaturated fatty acids, 145
Potassium, 56, 92—94, 132, 142, 160
    calcium-ATPase inhibition and, 154
    calcium-dependent increase in permeability of,
        37
    calcium efflux by, 134
    of calcium pump, 92
    magnesium and, 93
Potassium channel, 38
Praseodymium, 154
Presynaptic nerve terminals, 6, 8
Properties
    of calcium pump, 51—56
    of purified calcium-ATPase, 62—64
Protein kinases, 29, 31—34
Protein-phosphatases, 30—31
Proteins, see also specific proteins, 6
    activators of, 148
    calcium-binding, see Calcium-binding proteins
    compounds that react with, 159—161
    contractile, 35
    inhibitors of, 148
Proteolysis, 59, 62, 74, 93, 146—148
    partial, 77
Purification
    of calcium-ATPase, 60—62
    of calcium pump, 59—65
Purified calcium-ATPase, 62—64, 140

Q

Quercetin, 162

Quin2, see 2,2-Bis(ethoxycarboxyl)methylamino-
    5-methylphenoxymethyl-6-methoxy-8-
    bis(ethoxycarbonyl)methylaminoquinoline
Quinidine, 38
Quinine, 38

R

Rate constants for calmodulin binding, 141
Rate equations, 85—86
Reaction mechanism of calcium-ATPase, 83
Reaction scheme for ATP hydrolysis, 114—115
Receptor-mediated calcium mobilization, 24
Reconstitution
    of calcium pump, 60, 69, 76
    of ghosts of red blood cells, 67—69, 71
    of liposomes, 67, 69—70
    of purified calcium-ATPase, 64
Red blood cells, 6, 8, 17, 18, 56, 98
    calcium efflux from intact, 67
    ghosts of, see Ghosts of red blood cells
    influx of calcium in, 14
    intact, 67—68
Regulation
    of calcium concentration in cytosol, 16—19
    of calcium pump, 137—149
    of cAMP, 31
    of cell functions by calcium, 34—39
    of motile processes, 31
    of muscle contraction, 31
    by phosphorylation, 148—149
Regulatory domain, 62, 83, 148
Release
    of calcium from intracellular stores, 24
    of inhibition by high calcium, 156
    of neurotransmitters, 34
Resealed ghosts of red blood cells, 67—69, 71
Reservoir of calcium, 6
Residual (uncoupled) efflux of calcium, 14
Resting cells, 17—18
Reticulum
    endoplasmic, 8, 15, 17, 24, 95
    sarcoplasmic, see Sarcoplasmic reticulum
Reversal
    of calcium pump, 101—102
    of phosphorylation, 107
Reversible hemolysis, 68
Rubidium, 132
Ruthenium red, 162

S

Salicyclic acid, 93
Samarium, 154
Sarcolemma, 48, 53, 56, 62
Sarcoplasmic reticulum, 6, 8, 59, 98, 101, 156,
    162
    ATP hydrolysis and, 81
    homeostasis and, 16, 17

partial reactions and, 108, 116, 122
protein kinases of, 31
of skeletal muscle, 31
transport and, 79
Second messenger role of calcium, 21
Self-assembling cell components, 35—36
Sensitivity of calmodulin, 62
Sepharose CL-6B, 59
Shell gland, 54, 56
Sialic acid, 7
Sideness of activation, 134
Sinapsis alba (mustard), 48
Single-channel conductances, 23
Sites for ATP, 113
Sites for calcium, 94—96
Skeletal muscle, 6, 9, 24
    phosphorylase kinase of, 29—30
Smooth muscle, 6, 9, 56, 62
Sodium, 132
    calcium efflux by, 134
Sodium-calcium countertransport, 47
Sodium-calcium exchange, 13, 14
Sodium, potassium-ATPase, 108, 120, 162
Sodium pump, 81, 122
    ATPases of, 59
Solubilization, 59
Sonication, 69
Specific activity of purified enzyme, 61, 63
Specificity
    calcium, 91
    substrate, 80—81
Specific requirements for ATP, 80
Spectrin, 35
Sperm, 55, 56
Squash (*Cucurbita pepo*), 48
Squid axons, 9, 13, 17, 52, 67, 70—72
    internally dialyzed, 47
Stability of purified calcium-ATPase, 62
Steady-state conditions, 129
Steady-state kinetics of calcium-ATPase, 142—
    144
Steady-state level of phosphorylation, 107
Stereoisomer EF-hand relationships, 25
Stimulated cells, 21—24
Stoichiometrical coupling between ATP hydroly-
    sis and calcium transport, 50
Stoichiometry
    for binding of calmodulin to calcium-ATPase,
    139
    of calcium pump, 101
    of calcium transport, 99—101
Stomach muscle, 53, 56
Storage of calcium, 6
Striated muscle, 8
Strontium, 91
Substrate, see also specific types
    as activator, 84—85
    ATP as, 81, 87
    energy-donating, 81
    ITP as, 81
    kinetics of, 156

magnesium-ATP as, 131
specificity of, 80—81
two sites for, 83
UTP as, 81
Substrate curves, 81—83, 116—117
    biphasic, 83—86, 113
    of calcium-ATPase, 143
    kinetic analysis of, 83—86
Sugars, 6
Surface density
    of calcium pump, 107
    of channels, 23
Synapsin I, 32
Synthesis of ATP by calcium pump, 102

T

Target size, 62
Tau factor, 36
Temperature dependence of calcium, 98
Tetracarboxylate dyes, 2, 3
Tetracarboxylic acids, 3
Tetradecanoylphorbol (TPA), 33, 34
Tetraphenylboron, 76
TFP, see Trifluperazine
Tissues, see also specific types
    calmodulin concentration in, 27
    cells of, 53—54
Titration, 5
Total calcium, 1
TPA, see Tetradecanoylphorbol
Transients in calcium concentration, 5
Translocation of hydrogen, 76
Transport
    in artificial lipid membranes, 64
    by external calcium, 79
    in regulation of calcium concentration in cyto-
        sol, 16—19
Transport of calcium, 67—79
    activation energy for, 97
    calmodulin effect on, 142
    charge balance during, 76
    coupling between ATP hydrolysis and, 50, 99—
        102
    coupling of ATP hydrolysis and, 91—102
    dependence on pH, 96
    electrical balance during, 75—79
    electrogenic, 75—76
    electroneutral, 76—77
    energetics of, 99
    hydrogen translocation during, 76
    identification of, 49—51
    increase in rate of, 74—75
    inhibition of, 73, 74
    by intracellular organelles, 15—16
    kinetic parameters of, 17—18
    maximum rate of, 72, 74
    membrane preparations for study of, 67—71
    phosphatase activity and, 122—123
    in resealed ghosts of red blood cells, 73

sodium efflux and, 135
stoichiometrical coupling between ATP hydrolysis and, 50
stoichiometry of, 99—101
Trifluoperazine (TFP), 93, 156
Triton X-100®, 59, 95
Troponin, 26, 30, 138
Trypsin, 62, 146
Tryptophan-5-monoxygenase, 32
Tubulin, 36
Turnover
    of calcium ATPase, 63
    of phosphoenzyme, 144
Tween-20®, 59

U

Uncoupled (residual) efflux of calcium, 14
Uptake of calcium, 14
    by inside-out vesicles, 73
    mitochondrial, 15
UTP, 80, 119
    as substrate, 81

V

Valinomycin, 76
Vanadate, 154—156
Vanadium, 154
Vesicles, see also specific types
    inside-out, see Inside-out vesicles (IOV)
    sarcolemma, 48
    volume of, 69
Vitamin D3, 26
Vitamin D-dependent calcium-binding proteins, 26
Voltage-sensitive channels for calcium, 13
Volume of vesicles, 69

X

X-ray analysis, 6

Z

Zea mays (corn), 48, 54, 56

Printed and bound by CPI Group (UK) Ltd, Croydon, CR0 4YY

22/10/2024

01777633-0020